MATH SKILLBUILDER

gold

Columbus, OH

SRAonline.com

SRA

Printed in the United States of America.

Send all inquiries to this address:
SRA/McGraw-Hill
8787 Orion Place
Columbus, OH 43240-4027

ISBN: 978-0-07-618616-7
MHID: 0-07-618616-4

6 7 8 9 10 11 12 GLO 21 20 19 18 17

The McGraw-Hill Companies

Contents

6 CHAPTER 6

Measurement

7 CHAPTER 7

Geometry

8 CHAPTER 8

Addition and Subtraction (2-digit with no renaming)

9 CHAPTER 9

Addition and Subtraction (facts through 18)

PROBLEM SOLVING STRATEGIES

Find the Facts

There are 4  .

There are 2 .

What are the facts?

How many in all?

Ring the facts.

There are 4 .

There are 2 .

How many in all?

Ring the facts.

I have 5 little  .

I have 2 big  .

How many in all?

There are 6  .

4 fly away.

How many are left?

Cari has 3 .

Angie has 3 .

How many in all?

NAME ______________________

PROBLEM SOLVING STRATEGIES

Find the Question

There are 4 . What is the question?

2 roll away.

How many are left?

Ring the question.

There are 4 .

3 roll away.

How many are left?

Ring the question.

Amy has 7 .

Mica has 2 more.

How many in all?

There were 5 .

Then 4 more came.

Now how many in all?

There were 4 .

2 flew away.

How many are left?

PROBLEM SOLVING STRATEGIES

Choose the Correct Operation

Sara saw 4 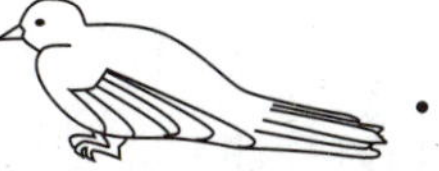.

Kyle saw 3 more.

How many in all?

There are __7__ birds.

Do you add or subtract?

To find out how many, **add.**

$$\begin{array}{r} 4 \\ +\ 3 \\ \hline 7 \end{array}$$

Ring the correct operation. Then solve the problem.

There are 7 .

2 run away.

How many are left?

add or subtract

____ dogs are left.

SHOW YOUR WORK

There are 5 little .

There are 4 big .

How many in all?

add or subtract

There are ____ books.

PROBLEM SOLVING STRATEGIES

Draw a Picture

I saw 5 .

Barry saw 2 more.

How many in all?

There are __7__ in all.

What picture can you draw?

Draw a picture to add.

Draw a picture. Then solve the problem.

There are 6 .

3 drive away.

How many are left?

There are ____ left.

Jodi saw 4 .

Jarod saw 5 .

How many in all?

There are ____ in all.

NAME ______________________

PROBLEM SOLVING STRATEGIES

Look for a Pattern

What is the pattern?

Ring what comes next.

 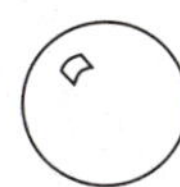  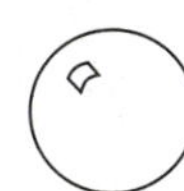  

Ring what comes next.

PROBLEM SOLVING STRATEGIES

Guess and Check

Ann has 2 pencils.

Tim has 4 pencils.

Kyle has 5 pencils.

Which two have 6 pencils in all?

(Ann and Tim) Tim and Kyle

Ann and Kyle

Guess: Ann and Tim

Check:

$$\begin{array}{r} 2 \\ + \ 4 \\ \hline 6 \end{array}$$

Solve each problem. Then ring the answer.

Anna has 3 apples.

Tom has 2 apples.

Jen has 1 apple.

Which two have 4 apples in all?

Anna and Tom Tom and Jen

Anna and Jen

SHOW YOUR WORK

Guess:

Check:

Ed has 8 .

Kim has 2 .

Taylor has 1 .

Which two have 10 balloons in all?

Ed and Kim Kim and Taylor

Ed and Taylor

Guess:

Check:

Lesson 1 Numbers 0 through 3

zero	one	two	three
0	1	2	3

Ring the numeral.

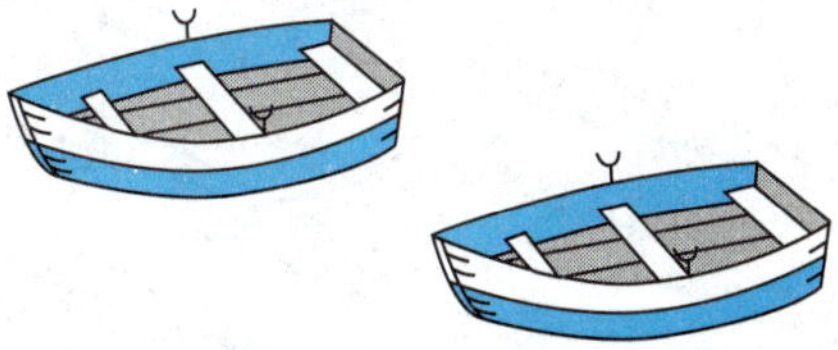

0 1 2 3

0 1 2 3

0 1 2 3

0 1 2 3

0 1 2 3

0 1 2 3

Lesson 2 Numbers 4 through 7

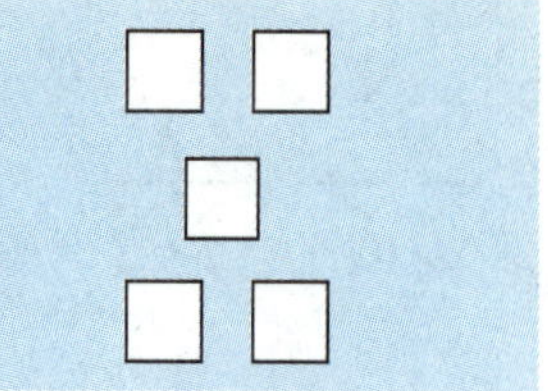

four	five	six	seven
4	5	6	7

Ring the numeral.

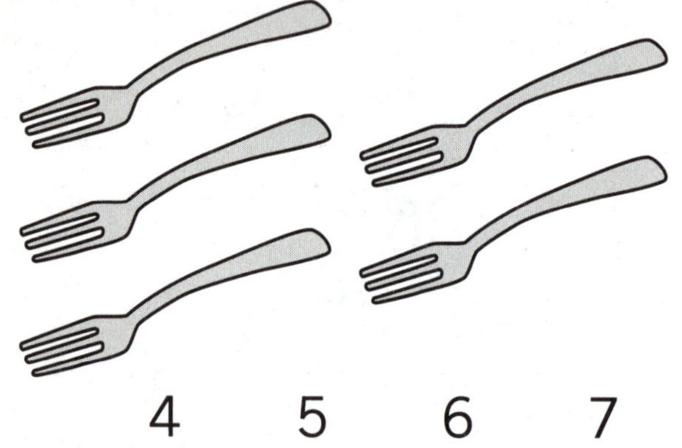

4 5 6 7

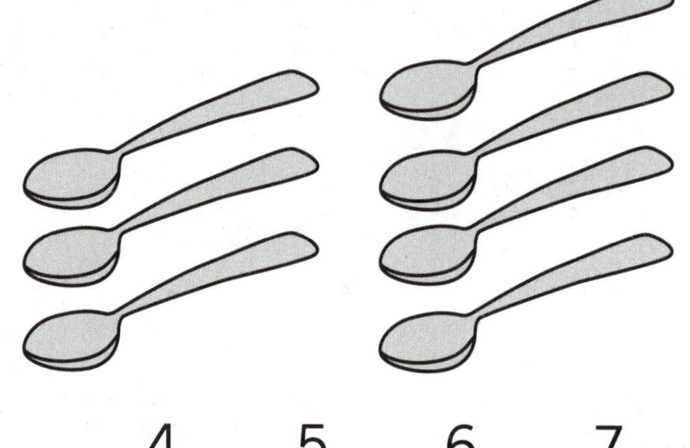

4 5 6 7

4 5 6 7

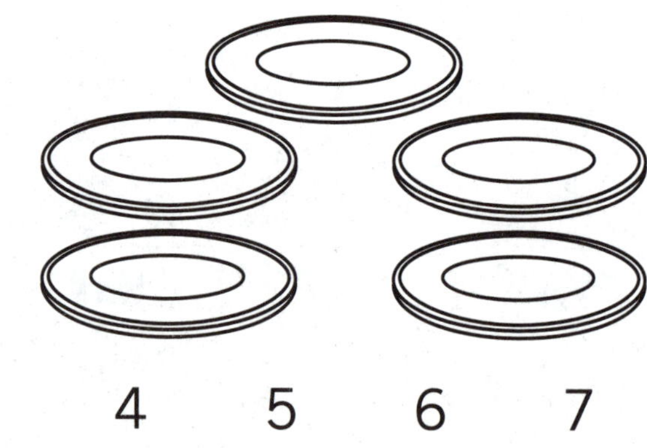

4 5 6 7

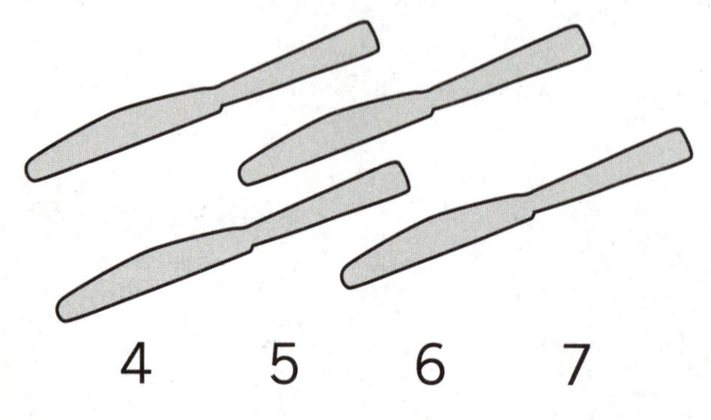

4 5 6 7

4 5 6 7

Lesson 3 Numbers 8 through 10

eight	nine	ten
8	9	10

Ring the numeral.

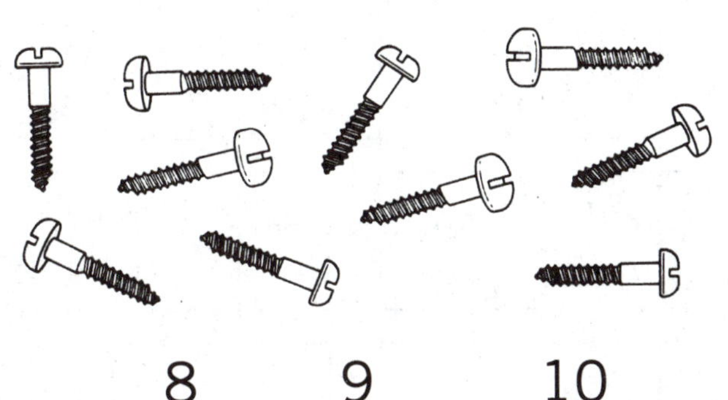

8 9 10

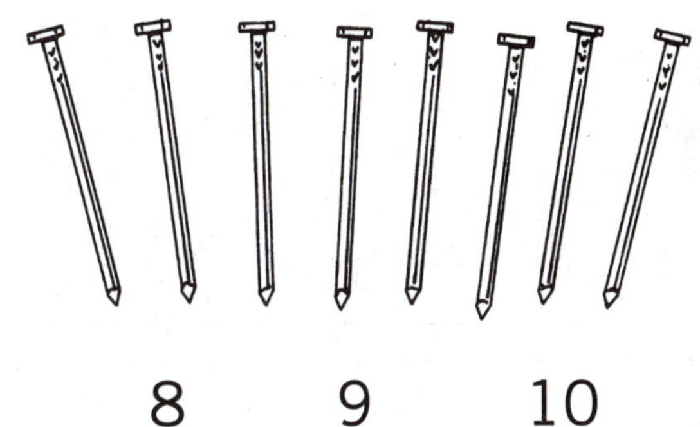

8 9 10

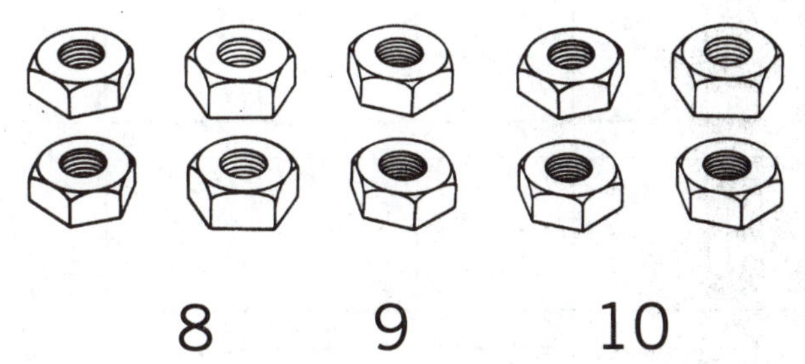

8 9 10

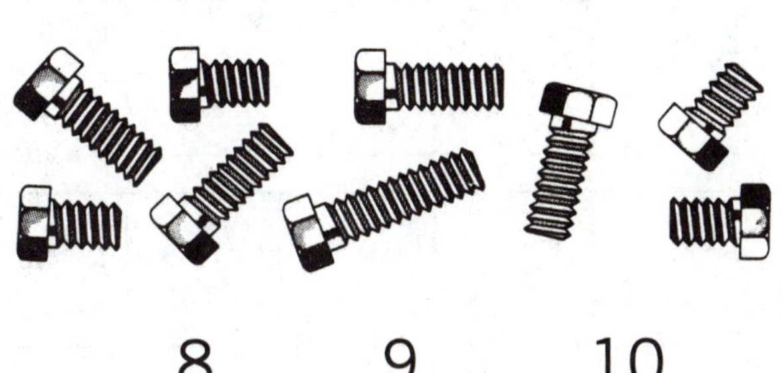

8 9 10

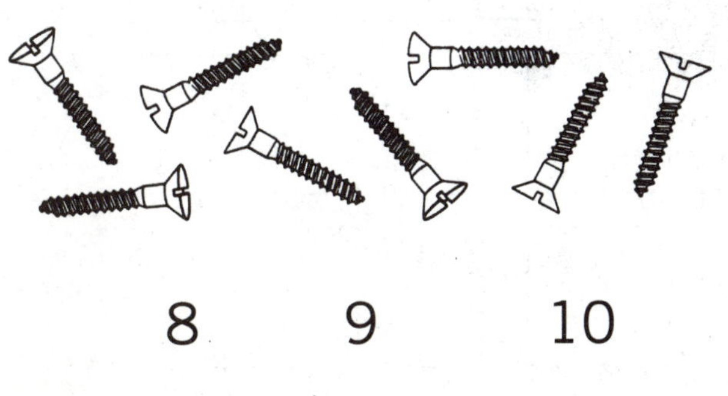

8 9 10

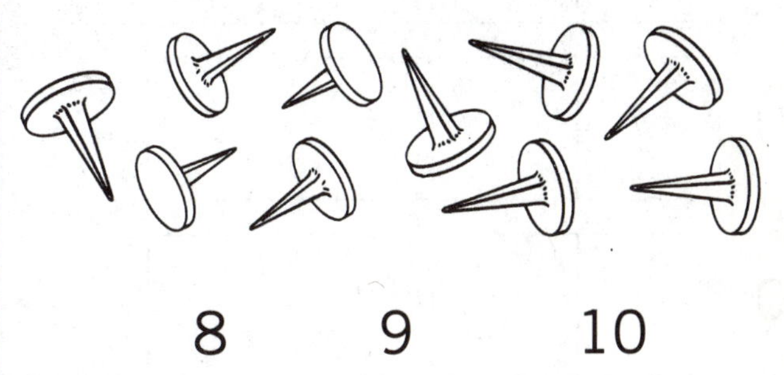

8 9 10

NAME ______________________________

Lesson 4 Numbers 0 through 10

Read the numeral. Color that many squares.

0										
1										
2										
3										
4										
5										
6										
7										
8										
9										
10										

NAME ______________________

Lesson 5 Numbers 0 through 10

Tell how many.

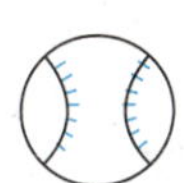
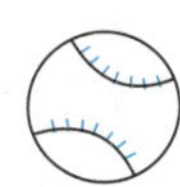
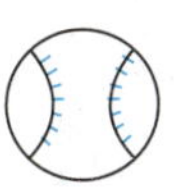
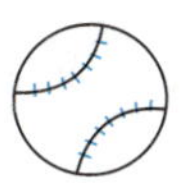
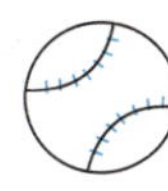
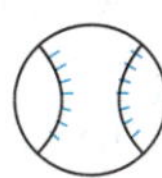
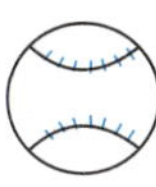

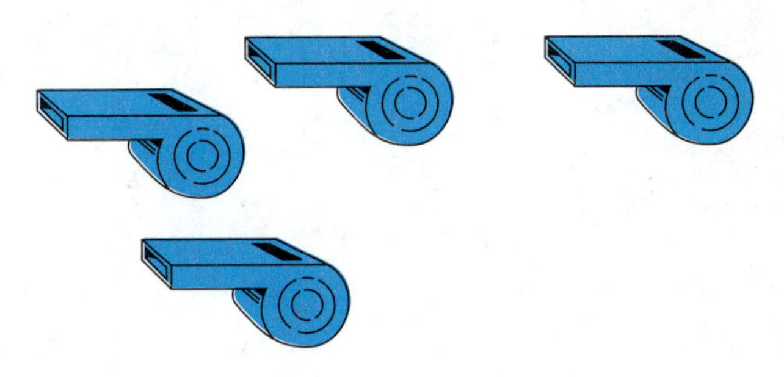

______

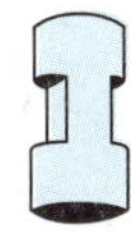
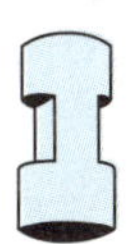
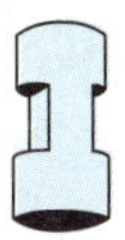

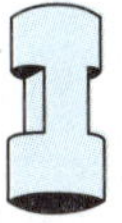

______

______

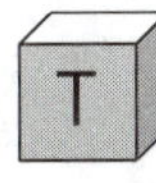

______

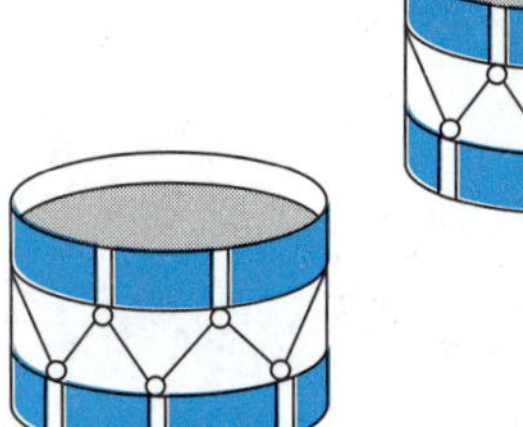

______

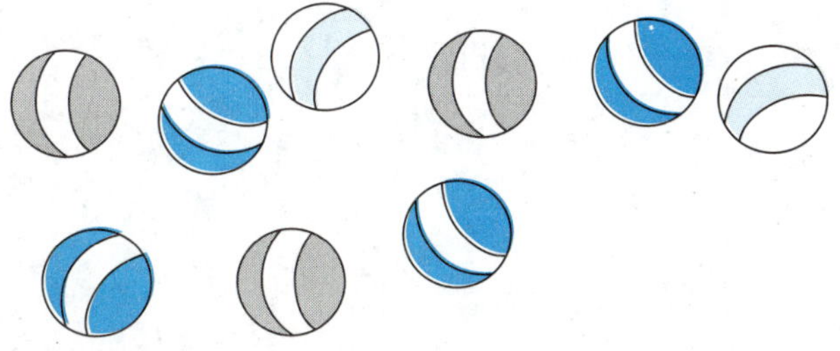

______

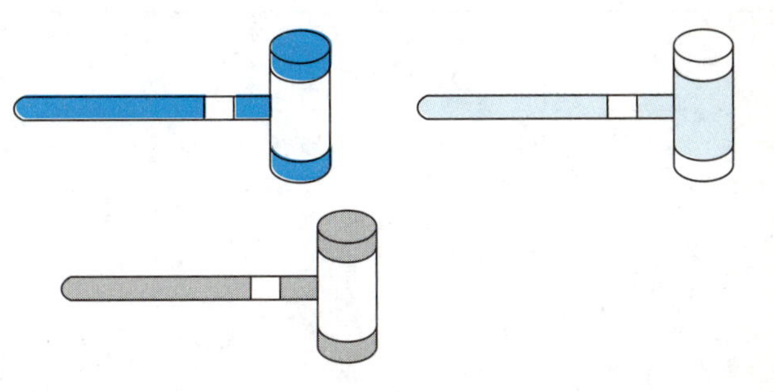

______

CHAPTER 1

Lesson 5 Numbers 0 through 10

Read each name for the number.
Then draw that many ●s.

three 3	ten 10
five 5	seven 7
two 2	eight 8
one 1	nine 9

Lesson 6 Order of Numbers

Fill in each 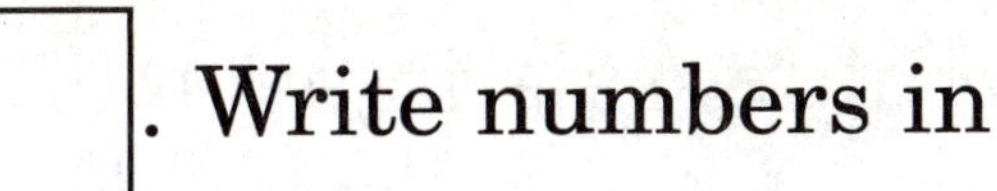

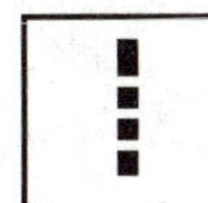  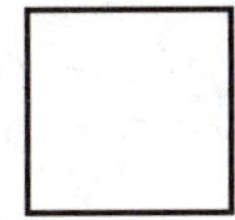  

 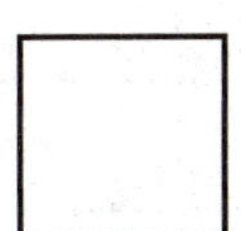 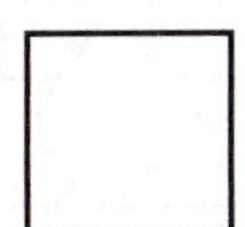 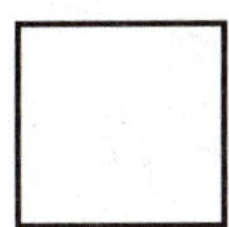 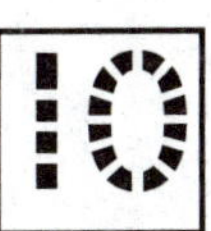

Count backward. Write a numeral on each ______.

10 ______ ______ ______ 6

5 ______ ______ ______ 1

Count forward. Write a numeral in each ☐.

Lesson 6 Order of Numbers

Count forward. Write the numerals in each ▭.
Then trace the number words on the ≡.

Lesson 7 Order of Numbers

Connect the dots in order.

Start here.

0 2 4 6

10

1 3 5

9 8 7

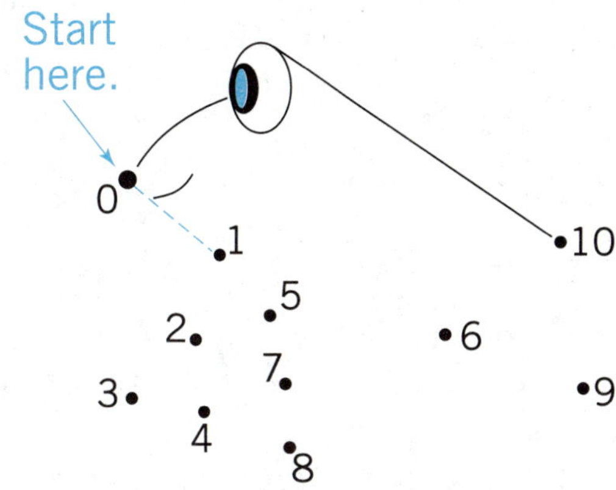

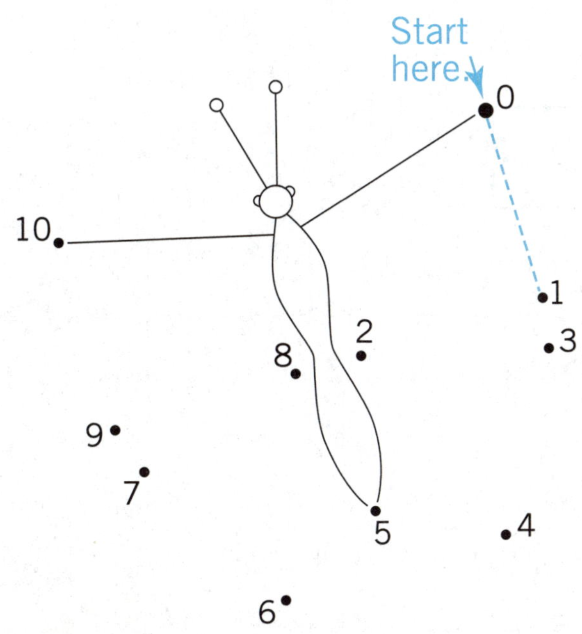

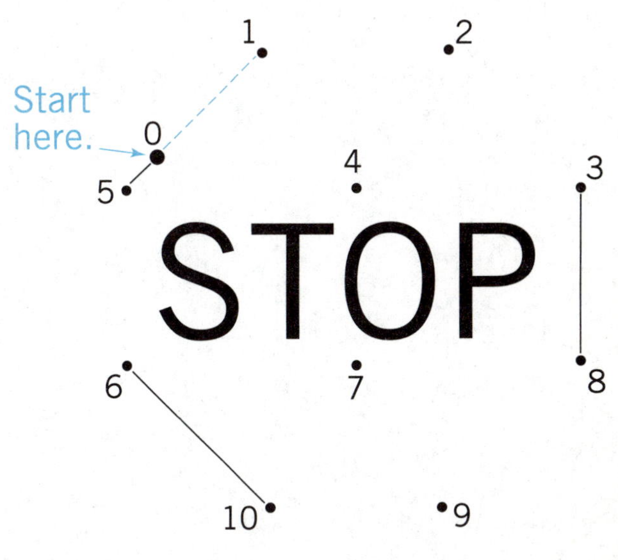

NAME ________________

Lesson 8 Number Patterns

Finish the pattern. Write a number in each ☐.

3	4	☐	6	7
2	☐	4	5	6
6	7	☐	9	10
6	5	4	3	☐
8	7	☐	5	4

Lesson 9 Ordinal Numbers

Ring the third .

Ring the fifth 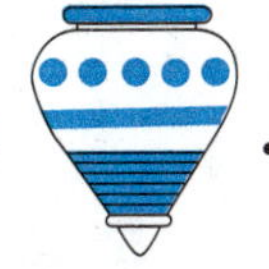.

 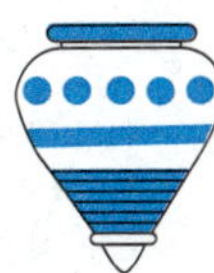   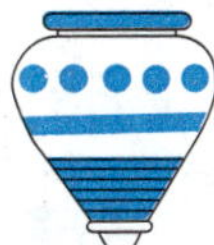  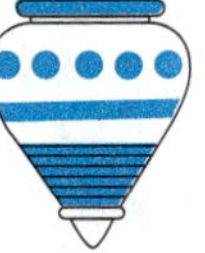

Ring the second .

Ring the eighth .

   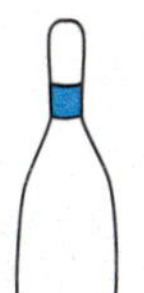  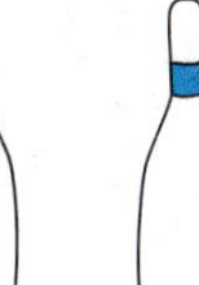  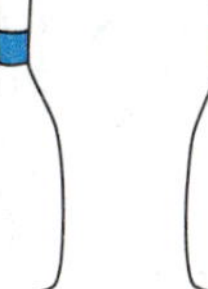  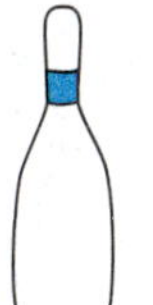  

Ring the fourth 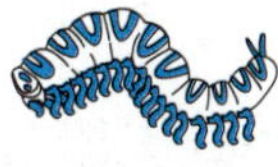.

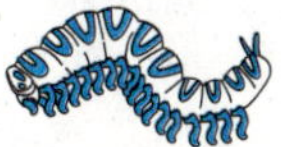  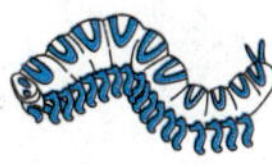 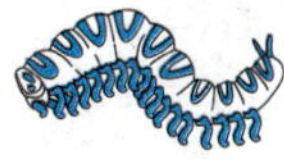 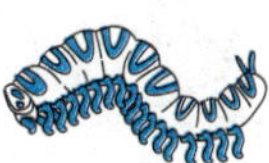 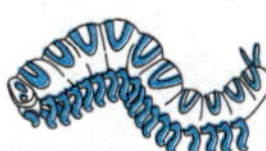

Lesson 9 Ordinal Numbers

Ring the ninth.

Ring the first 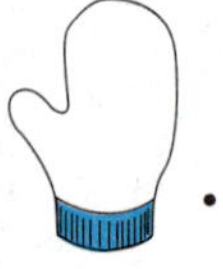.

 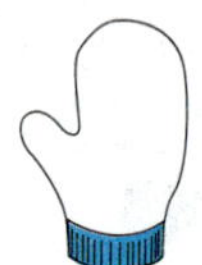      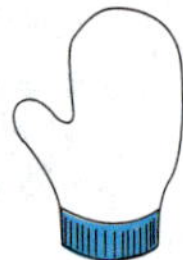

Ring the sixth .

Ring the tenth 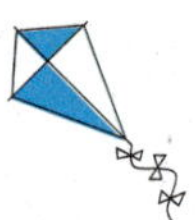.

 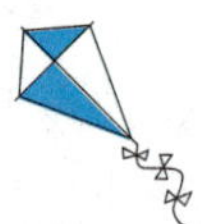   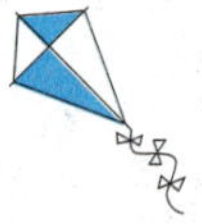 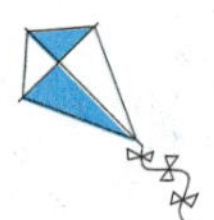  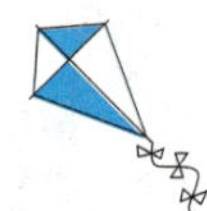 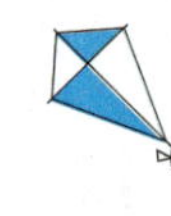 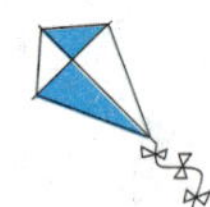

Ring the seventh 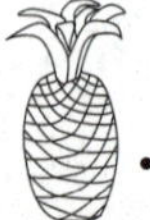.

     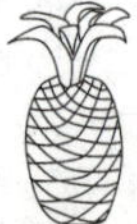 

NAME ______________________

CHAPTER 1 PRACTICE TEST
Numeration (0 through 10)

CHAPTER 1

Tell how many.

Read the numeral. Draw that many ◯s.

10	2	8

Write a numeral in each ☐ to finish the pattern.

2 3 ☐ 5 6

6 ☐ 8 9 10

Ring the third .

NAME ______________________

CHAPTER 2 PRETEST

Addition and Subtraction (facts through 5)

Add.

1 +2	0 +5	4 +1	2 +2	1 +0	2 +1
0 +2	4 +0	1 +3	3 +1	1 +4	3 +2
2 +0	0 +1	2 +3	5 +0	0 +4	2 +1

Subtract.

4 −1	5 −3	4 −2	5 −0	5 −1	3 −1
2 −0	5 −4	3 −2	3 −3	5 −2	4 −3
5 −5	4 −0	2 −2	3 −1	2 −1	4 −4

Lesson 1 Sums through 3

Add.

$\begin{array}{r} 1 \\ +1 \\ \hline 2 \end{array}$

1 + 1 = 2

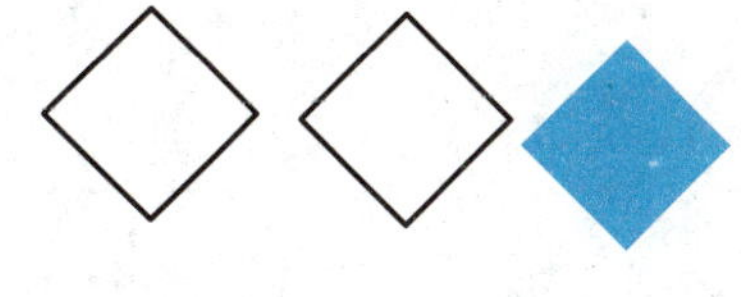

$\begin{array}{r} 2 \\ +1 \\ \hline \end{array}$

2 + 1 = ____

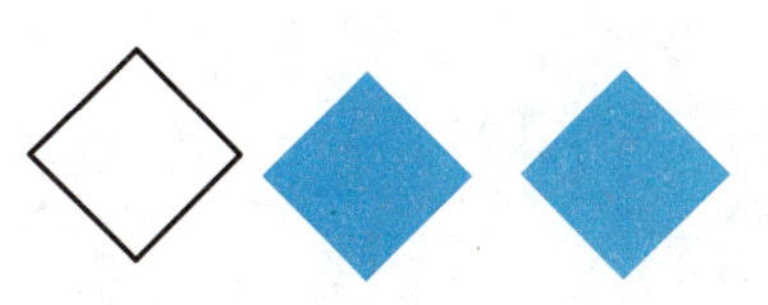

$\begin{array}{r} 1 \\ +2 \\ \hline \end{array}$

1 + 2 = ____

$\begin{array}{r} 2 \\ +0 \\ \hline \end{array}$

2 + 0 = ____

$\begin{array}{r} 0 \\ +2 \\ \hline \end{array}$

0 + 2 = ____

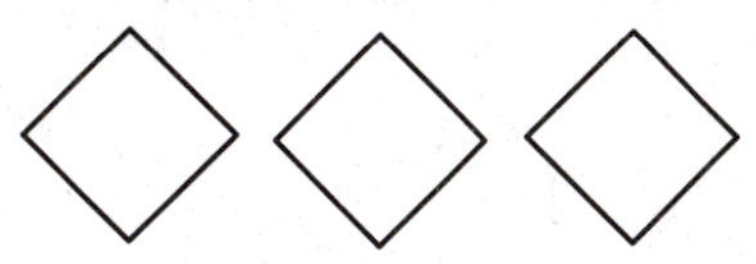

$\begin{array}{r} 3 \\ +0 \\ \hline \end{array}$

3 + 0 = ____

$\begin{array}{r} 0 \\ +3 \\ \hline \end{array}$

0 + 3 = ____

$\begin{array}{r} 0 \\ +0 \\ \hline \end{array}$

0 + 0 = ____

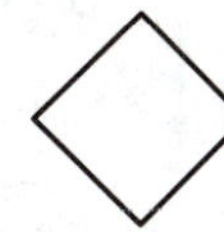

$\begin{array}{r} 1 \\ +0 \\ \hline \end{array}$

1 + 0 = ____

$\begin{array}{r} 0 \\ +1 \\ \hline \end{array}$

0 + 1 = ____

NAME ______________________

Lesson 2 Sums of 4 and 5

Add.

$\begin{array}{r} 4 \\ +1 \\ \hline 5 \end{array}$

4 + 1 = 5

$\begin{array}{r} 2 \\ +3 \\ \hline \end{array}$

2 + 3 = ____

$\begin{array}{r} 1 \\ +4 \\ \hline \end{array}$

1 + 4 = ____

$\begin{array}{r} 3 \\ +2 \\ \hline \end{array}$

3 + 2 = ____

$\begin{array}{r} 2 \\ +2 \\ \hline \end{array}$

2 + 2 = ____

$\begin{array}{r} 4 \\ +0 \\ \hline \end{array}$

4 + 0 = ____

$\begin{array}{r} 0 \\ +4 \\ \hline \end{array}$

0 + 4 = ____

$\begin{array}{r} 0 \\ +5 \\ \hline \end{array}$

0 + 5 = ____

$\begin{array}{r} 1 \\ +3 \\ \hline \end{array}$

1 + 3 = ____

$\begin{array}{r} 5 \\ +0 \\ \hline \end{array}$

5 + 0 = ____

$\begin{array}{r} 3 \\ +1 \\ \hline \end{array}$

3 + 1 = ____

NAME ______________________

Lesson 3 Subtracting from 1, 2, and 3

Subtract.

$$\begin{array}{r} 3 \\ -1 \\ \hline 2 \end{array}$$

3 − 1 = 2

$$\begin{array}{r} 2 \\ -1 \\ \hline \end{array}$$

2 − 1 = ______

$$\begin{array}{r} 3 \\ -2 \\ \hline \end{array}$$

3 − 2 = ______

$$\begin{array}{r} 1 \\ -0 \\ \hline \end{array}$$

1 − 0 = ______

$$\begin{array}{r} 3 \\ -0 \\ \hline \end{array}$$

3 − 0 = ______

$$\begin{array}{r} 1 \\ -1 \\ \hline \end{array}$$

1 − 1 = ______

$$\begin{array}{r} 2 \\ -2 \\ \hline \end{array}$$

2 − 2 = ______

$$\begin{array}{r} 3 \\ -3 \\ \hline \end{array}$$

3 − 3 = ______

NAME ______________________

Lesson 4 Subtracting from 4 and 5

Subtract.

$\begin{array}{r} 5 \\ -1 \\ \hline 4 \end{array}$

5 − 1 = 4

$\begin{array}{r} 4 \\ -3 \\ \hline \end{array}$

4 − 3 = ______

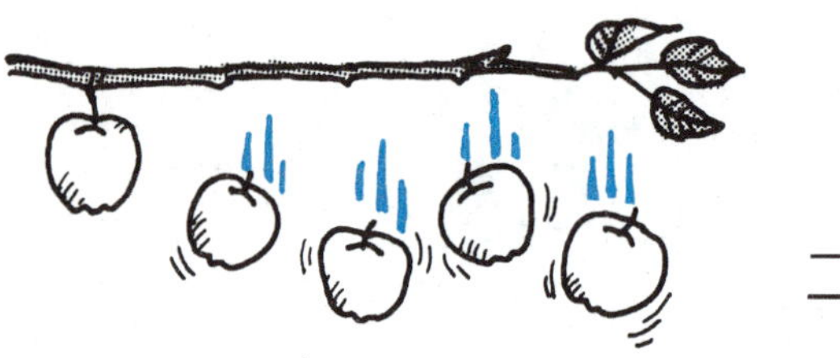

$\begin{array}{r} 5 \\ -4 \\ \hline \end{array}$

5 − 4 = ______

$\begin{array}{r} 4 \\ -4 \\ \hline \end{array}$

4 − 4 = ______

$\begin{array}{r} 5 \\ -2 \\ \hline \end{array}$

5 − 2 = ______

$\begin{array}{r} 4 \\ -2 \\ \hline \end{array}$

4 − 2 = ______

$\begin{array}{r} 5 \\ -3 \\ \hline \end{array}$

5 − 3 = ______

$\begin{array}{r} 4 \\ -1 \\ \hline \end{array}$

4 − 1 = ______

NAME ______________________

Lesson 5 Number Families

Add or subtract.

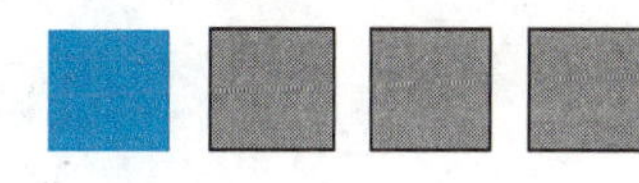

2	1	3	3	1	3	4	4
+1	+2	−2	−1	+3	+1	−1	−3
3	3	1					

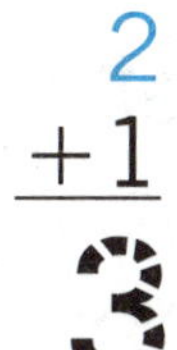

2	3	5	5	4	1	5	5
+3	+2	−2	−3	+1	+4	−4	−1

1	2	2	4
+1	−1	+2	−2

2	0	2	2	5	0	5	5
+0	+2	−2	−0	+0	+5	−5	−0

CHAPTER 2.

NAME ______________________

Lesson 6 Penny and Nickel

1 penny
1¢

5 pennies
5¢

1 nickel
5¢

Tell how much money.

5 ¢

___ ¢

___ ¢

___ ¢

___ ¢

___ ¢

___ ¢

___ ¢

Lesson 7 Problem Solving

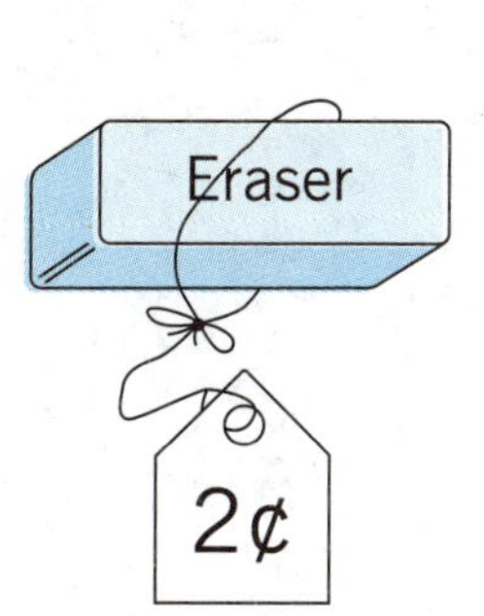

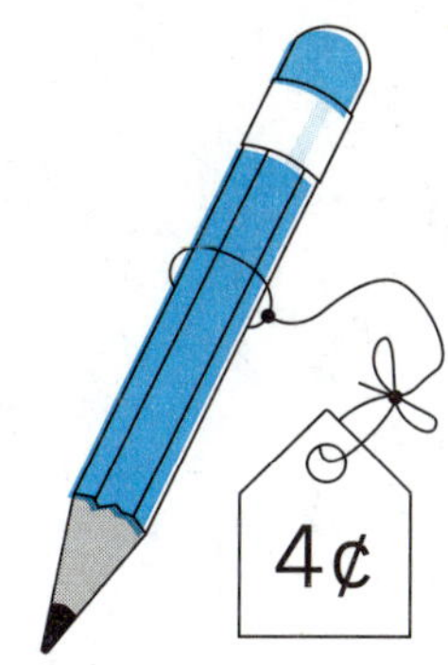

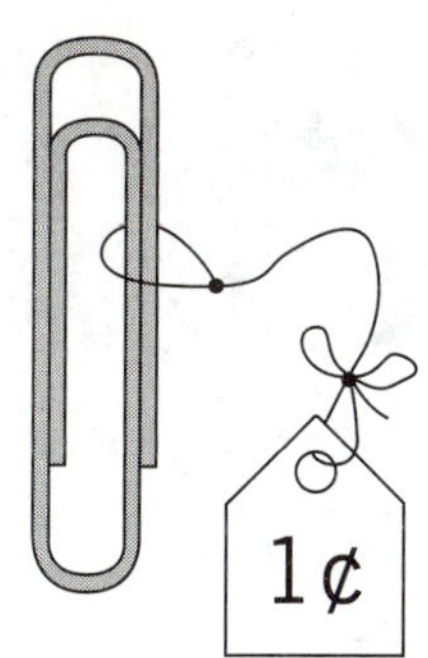

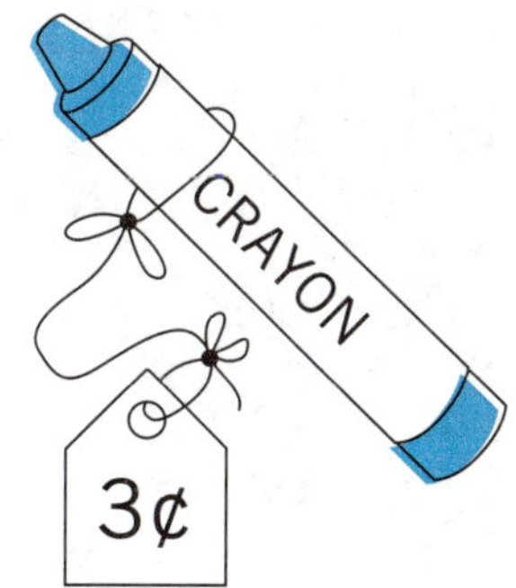

Add or subtract.

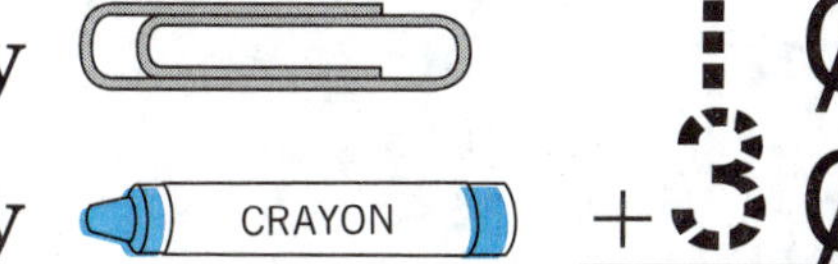

I buy 1¢

I buy + 3¢

I spend 4¢

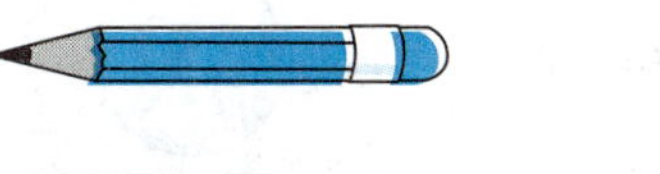

I buy ¢

I buy + ¢

I spend ¢

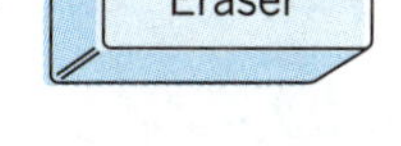

I buy ¢

I buy + ¢

I spend ¢

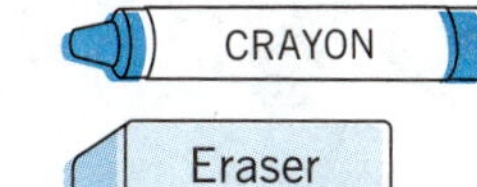

I buy ¢

I buy + ¢

I spend ¢

I have 5¢

I buy − 3¢

I have left ¢

I have ¢

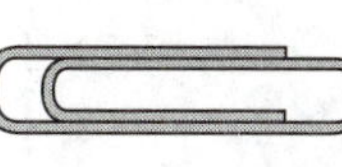

I buy − ¢

I have left ¢

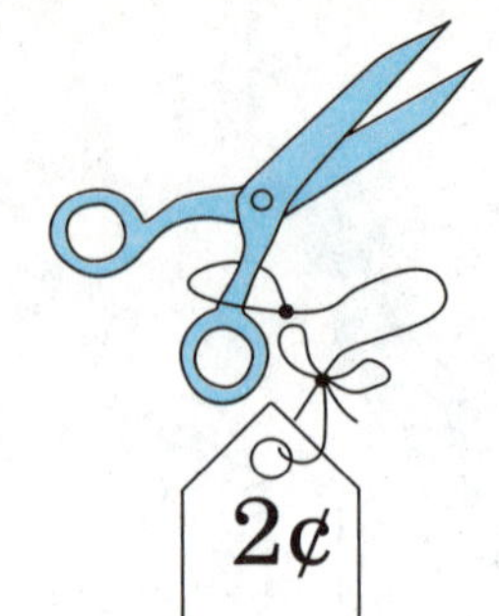

Add or subtract.

I have  5¢
I buy 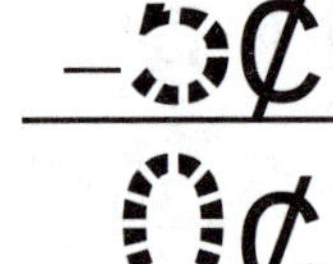 − 5¢
I have left 0¢

I buy  ¢
I buy  + ¢
I spend ¢

I have  ¢
I buy 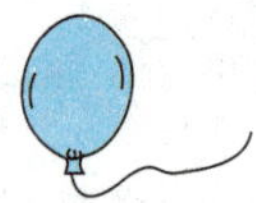 − ¢
I have left ¢

I buy  ¢
I buy  + ¢
I spend ¢

I have   ¢
I buy  − ¢
I have left ¢

I buy  ¢
I buy  + ¢
I spend ¢

CHAPTER 2 PRACTICE TEST

Addition and Subtraction (facts through 5)

Add.

3 +1	1 +1	0 +0	2 +1	5 +0	1 +0
0 +3	2 +2	1 +2	0 +1	1 +4	3 +2
1 +3	4 +0	0 +5	2 +0	4 +1	2 +3

Subtract.

2 −1	4 −1	5 −0	3 −2	5 −4	4 −4
4 −0	5 −1	0 −0	4 −2	3 −0	5 −2
2 −0	4 −3	5 −3	3 −1	3 −3	1 −1

NAME ______________________

CHAPTER 3

CHAPTER 3 PRETEST

Addition and Subtraction (facts through 8)

Add.

$\begin{array}{r} 5 \\ +\ 3 \\ \hline \end{array}$ $\begin{array}{r} 3 \\ +\ 4 \\ \hline \end{array}$ $\begin{array}{r} 6 \\ +\ 0 \\ \hline \end{array}$ $\begin{array}{r} 4 \\ +\ 4 \\ \hline \end{array}$ $\begin{array}{r} 1 \\ +\ 5 \\ \hline \end{array}$

$\begin{array}{r} 2 \\ +\ 6 \\ \hline \end{array}$ $\begin{array}{r} 1 \\ +\ 7 \\ \hline \end{array}$ $\begin{array}{r} 2 \\ +\ 4 \\ \hline \end{array}$ $\begin{array}{r} 3 \\ +\ 3 \\ \hline \end{array}$ $\begin{array}{r} 6 \\ +\ 1 \\ \hline \end{array}$

$\begin{array}{r} 7 \\ +\ 0 \\ \hline \end{array}$ $\begin{array}{r} 4 \\ +\ 3 \\ \hline \end{array}$ $\begin{array}{r} 5 \\ +\ 2 \\ \hline \end{array}$ $\begin{array}{r} 2 \\ +\ 5 \\ \hline \end{array}$ $\begin{array}{r} 6 \\ +\ 2 \\ \hline \end{array}$

Subtract.

$\begin{array}{r} 6 \\ -\ 3 \\ \hline \end{array}$ $\begin{array}{r} 8 \\ -\ 4 \\ \hline \end{array}$ $\begin{array}{r} 7 \\ -\ 4 \\ \hline \end{array}$ $\begin{array}{r} 6 \\ -\ 0 \\ \hline \end{array}$ $\begin{array}{r} 7 \\ -\ 1 \\ \hline \end{array}$

$\begin{array}{r} 8 \\ -\ 6 \\ \hline \end{array}$ $\begin{array}{r} 6 \\ -\ 5 \\ \hline \end{array}$ $\begin{array}{r} 8 \\ -\ 3 \\ \hline \end{array}$ $\begin{array}{r} 6 \\ -\ 4 \\ \hline \end{array}$ $\begin{array}{r} 8 \\ -\ 7 \\ \hline \end{array}$

$\begin{array}{r} 7 \\ -\ 0 \\ \hline \end{array}$ $\begin{array}{r} 7 \\ -\ 3 \\ \hline \end{array}$ $\begin{array}{r} 7 \\ -\ 2 \\ \hline \end{array}$ $\begin{array}{r} 7 \\ -\ 5 \\ \hline \end{array}$ $\begin{array}{r} 8 \\ -\ 2 \\ \hline \end{array}$

NAME ________________

Lesson 1 Sums of 6

Add.

$$\begin{array}{r} 1 \\ +5 \\ \hline \end{array}$$

1 + 5 = 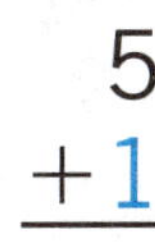

$$\begin{array}{r} 5 \\ +1 \\ \hline \end{array}$$

5 + 1 = ____

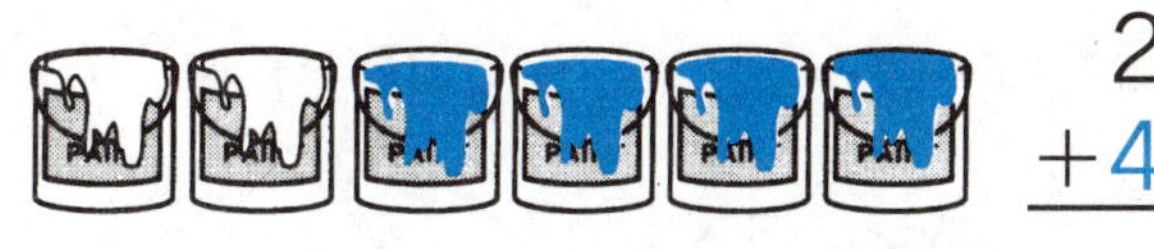

$$\begin{array}{r} 2 \\ +4 \\ \hline \end{array}$$

2 + 4 = ____

$$\begin{array}{r} 4 \\ +2 \\ \hline \end{array}$$

4 + 2 = ____

$$\begin{array}{r} 6 \\ +0 \\ \hline \end{array}$$

6 + 0 = ____

$$\begin{array}{r} 0 \\ +6 \\ \hline \end{array}$$

0 + 6 = ____

$$\begin{array}{r} 3 \\ +3 \\ \hline \end{array}$$

3 + 3 = ____

$\begin{array}{r} 2 \\ +4 \\ \hline \end{array}$	$\begin{array}{r} 1 \\ +3 \\ \hline \end{array}$	$\begin{array}{r} 5 \\ +1 \\ \hline \end{array}$	$\begin{array}{r} 3 \\ +3 \\ \hline \end{array}$	$\begin{array}{r} 2 \\ +2 \\ \hline \end{array}$	$\begin{array}{r} 3 \\ +2 \\ \hline \end{array}$

NAME ________________

Lesson 2 Subtracting from 6

Subtract.

$\begin{array}{r} 6 \\ -1 \\ \hline 5 \end{array}$

6 − 1 = 5

$\begin{array}{r} 6 \\ -5 \\ \hline \end{array}$

6 − 5 = ______

$\begin{array}{r} 6 \\ -4 \\ \hline \end{array}$

6 − 4 = ______

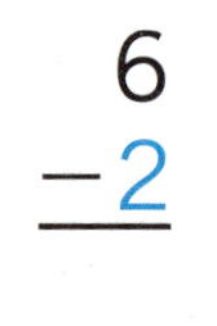

$\begin{array}{r} 6 \\ -2 \\ \hline \end{array}$

6 − 2 = ______

$\begin{array}{r} 6 \\ -3 \\ \hline \end{array}$

6 − 3 = ______

$\begin{array}{r} 6 \\ -0 \\ \hline \end{array}$

6 − 0 = ______

$\begin{array}{r} 6 \\ -3 \\ \hline \end{array}$ $\begin{array}{r} 6 \\ -6 \\ \hline \end{array}$ $\begin{array}{r} 6 \\ -1 \\ \hline \end{array}$ $\begin{array}{r} 6 \\ -4 \\ \hline \end{array}$ $\begin{array}{r} 6 \\ -2 \\ \hline \end{array}$ $\begin{array}{r} 6 \\ -5 \\ \hline \end{array}$

Lesson 3 Sums of 7

Add.

$\begin{array}{r} 3 \\ +4 \\ \hline 7 \end{array}$

3 + 4 = 7

$\begin{array}{r} 4 \\ +3 \\ \hline \end{array}$

4 + 3 = ____

$\begin{array}{r} 6 \\ +1 \\ \hline \end{array}$

6 + 1 = ____

$\begin{array}{r} 1 \\ +6 \\ \hline \end{array}$

1 + 6 = ____

$\begin{array}{r} 7 \\ +0 \\ \hline \end{array}$

7 + 0 = ____

$\begin{array}{r} 0 \\ +7 \\ \hline \end{array}$

0 + 7 = ____

$\begin{array}{r} 2 \\ +5 \\ \hline \end{array}$

2 + 5 = ____

$\begin{array}{r} 5 \\ +2 \\ \hline \end{array}$

5 + 2 = ____

$\begin{array}{r} 5 \\ +2 \\ \hline \end{array}$ $\begin{array}{r} 3 \\ +3 \\ \hline \end{array}$ $\begin{array}{r} 4 \\ +3 \\ \hline \end{array}$ $\begin{array}{r} 1 \\ +6 \\ \hline \end{array}$ $\begin{array}{r} 3 \\ +4 \\ \hline \end{array}$ $\begin{array}{r} 6 \\ +0 \\ \hline \end{array}$

NAME ____________________

Lesson 4 Subtracting from 7

Subtract.

$$\begin{array}{r} 7 \\ -6 \\ \hline 1 \end{array}$$

7 − 6 = 1

$$\begin{array}{r} 7 \\ -1 \\ \hline \end{array}$$

7 − 1 = ____

$$\begin{array}{r} 7 \\ -3 \\ \hline \end{array}$$

7 − 3 = ____

$$\begin{array}{r} 7 \\ -4 \\ \hline \end{array}$$

7 − 4 = ____

$$\begin{array}{r} 7 \\ -7 \\ \hline \end{array}$$

7 − 7 = ____

$$\begin{array}{r} 7 \\ -0 \\ \hline \end{array}$$

7 − 0 = ____

$$\begin{array}{r} 7 \\ -2 \\ \hline \end{array}$$

7 − 2 = ____

$$\begin{array}{r} 7 \\ -5 \\ \hline \end{array}$$

7 − 5 = ____

NAME ______________________

Lesson 5 Sums of 8

Add.

5
+3
8

5 + 3 = 8

3
+5

3 + 5 = ____

7
+1

7 + 1 = ____

1
+7

1 + 7 = ____

2
+6

2 + 6 = ____

6
+2

6 + 2 = ____

4
+4

4 + 4 = ____

CHAPTER 3

3	5	2	8	4	0
+3	+3	+6	+0	+3	+8

NAME ______________________

Lesson 6 Subtracting from 8

Subtract.

$$\begin{array}{r} 8 \\ -7 \\ \hline 1 \end{array}$$

$8 - 7 = 1$

$$\begin{array}{r} 8 \\ -1 \\ \hline \end{array}$$

$8 - 1 =$ ______

$$\begin{array}{r} 8 \\ -2 \\ \hline \end{array}$$

$8 - 2 =$ ______

$$\begin{array}{r} 8 \\ -6 \\ \hline \end{array}$$

$8 - 6 =$ ______

$$\begin{array}{r} 8 \\ -4 \\ \hline \end{array}$$

$8 - 4 =$ ______

$$\begin{array}{r} 8 \\ -8 \\ \hline \end{array}$$

$8 - 8 =$ ______

$$\begin{array}{r} 8 \\ -3 \\ \hline \end{array}$$

$8 - 3 =$ ______

$$\begin{array}{r} 8 \\ -5 \\ \hline \end{array}$$

$8 - 5 =$ ______

Lesson 7 Addition and Subtraction

Add.

5 +3	1 +6	3 +3	2 +6	1 +7	5 +1
0 +7	4 +2	4 +4	6 +0	3 +4	6 +1
3 +5	8 +0	5 +2	2 +4	6 +2	0 +8

Subtract.

8 −3	6 −5	8 −4	7 −3	6 −2	6 −6
8 −5	7 −0	8 −1	7 −6	8 −6	7 −5
8 −8	6 −3	7 −1	8 −2	6 −1	7 −2

Lesson 7 Problem Solving

Solve each problem.

There are 6 blue .

There are 2 white .

How many in all?

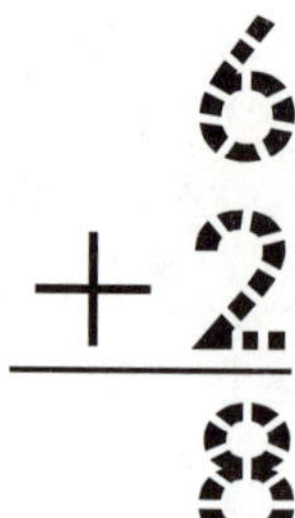

+

There are 7  .

4 fly away.

How many are left?

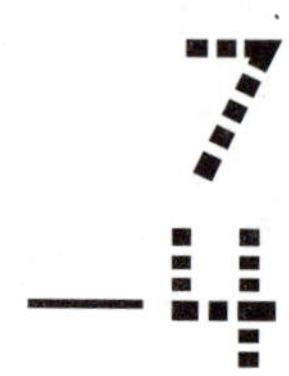

−

I saw 2 big  .

I saw 3 little  .

How many in all?

+

Bob has 8  .

Ann has 5  .

How many more does Bob have?

−

There is 1  .

Then 5 more come.

Now how many in all?

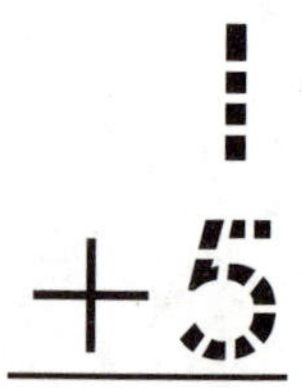

+

NAME ______________________

Lesson 8 Addition and Subtraction

Add or subtract. STOP! Watch the + and −.

$\begin{array}{r} 2 \\ +5 \\ \hline \end{array}$	$\begin{array}{r} 8 \\ -3 \\ \hline \end{array}$	$\begin{array}{r} 7 \\ +0 \\ \hline \end{array}$	$\begin{array}{r} 8 \\ -4 \\ \hline \end{array}$	$\begin{array}{r} 6 \\ +1 \\ \hline \end{array}$	$\begin{array}{r} 7 \\ -2 \\ \hline \end{array}$
$\begin{array}{r} 7 \\ -5 \\ \hline \end{array}$	$\begin{array}{r} 3 \\ +3 \\ \hline \end{array}$	$\begin{array}{r} 4 \\ +2 \\ \hline \end{array}$	$\begin{array}{r} 0 \\ +6 \\ \hline \end{array}$	$\begin{array}{r} 8 \\ -0 \\ \hline \end{array}$	$\begin{array}{r} 8 \\ -6 \\ \hline \end{array}$
$\begin{array}{r} 8 \\ -1 \\ \hline \end{array}$	$\begin{array}{r} 7 \\ -1 \\ \hline \end{array}$	$\begin{array}{r} 1 \\ +6 \\ \hline \end{array}$	$\begin{array}{r} 6 \\ -5 \\ \hline \end{array}$	$\begin{array}{r} 3 \\ +4 \\ \hline \end{array}$	$\begin{array}{r} 5 \\ +1 \\ \hline \end{array}$
$\begin{array}{r} 6 \\ -6 \\ \hline \end{array}$	$\begin{array}{r} 2 \\ +4 \\ \hline \end{array}$	$\begin{array}{r} 5 \\ +2 \\ \hline \end{array}$	$\begin{array}{r} 6 \\ -0 \\ \hline \end{array}$	$\begin{array}{r} 1 \\ +7 \\ \hline \end{array}$	$\begin{array}{r} 7 \\ -3 \\ \hline \end{array}$
$\begin{array}{r} 1 \\ +5 \\ \hline \end{array}$	$\begin{array}{r} 6 \\ -3 \\ \hline \end{array}$	$\begin{array}{r} 7 \\ -4 \\ \hline \end{array}$	$\begin{array}{r} 4 \\ +3 \\ \hline \end{array}$	$\begin{array}{r} 8 \\ -2 \\ \hline \end{array}$	$\begin{array}{r} 4 \\ +4 \\ \hline \end{array}$
$\begin{array}{r} 7 \\ -6 \\ \hline \end{array}$	$\begin{array}{r} 5 \\ +3 \\ \hline \end{array}$	$\begin{array}{r} 6 \\ -2 \\ \hline \end{array}$	$\begin{array}{r} 0 \\ +8 \\ \hline \end{array}$	$\begin{array}{r} 2 \\ +6 \\ \hline \end{array}$	$\begin{array}{r} 8 \\ -8 \\ \hline \end{array}$

Lesson 8 Problem Solving

Solve each problem.

There are 8 .

Then 7 ran away.

How many are left?

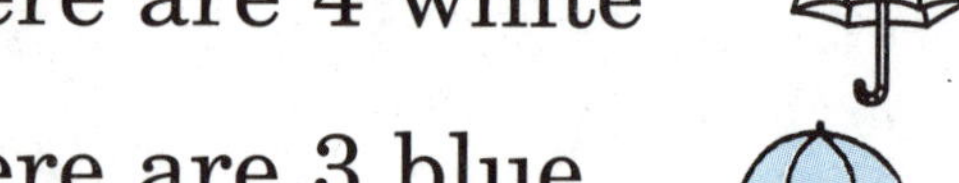
There are 4 white .

There are 3 blue .

How many in all?

Rob saw 7  flying.

He saw 1  sitting.

How many did he see in all?

Cal has 6  .

He gave Meg 4  .

How many does he have left?

There are 7 .

7 of the broke.

How many are left?

Lesson 9 Addition and Subtraction

Add or subtract.
If you get 6, color that part orange.

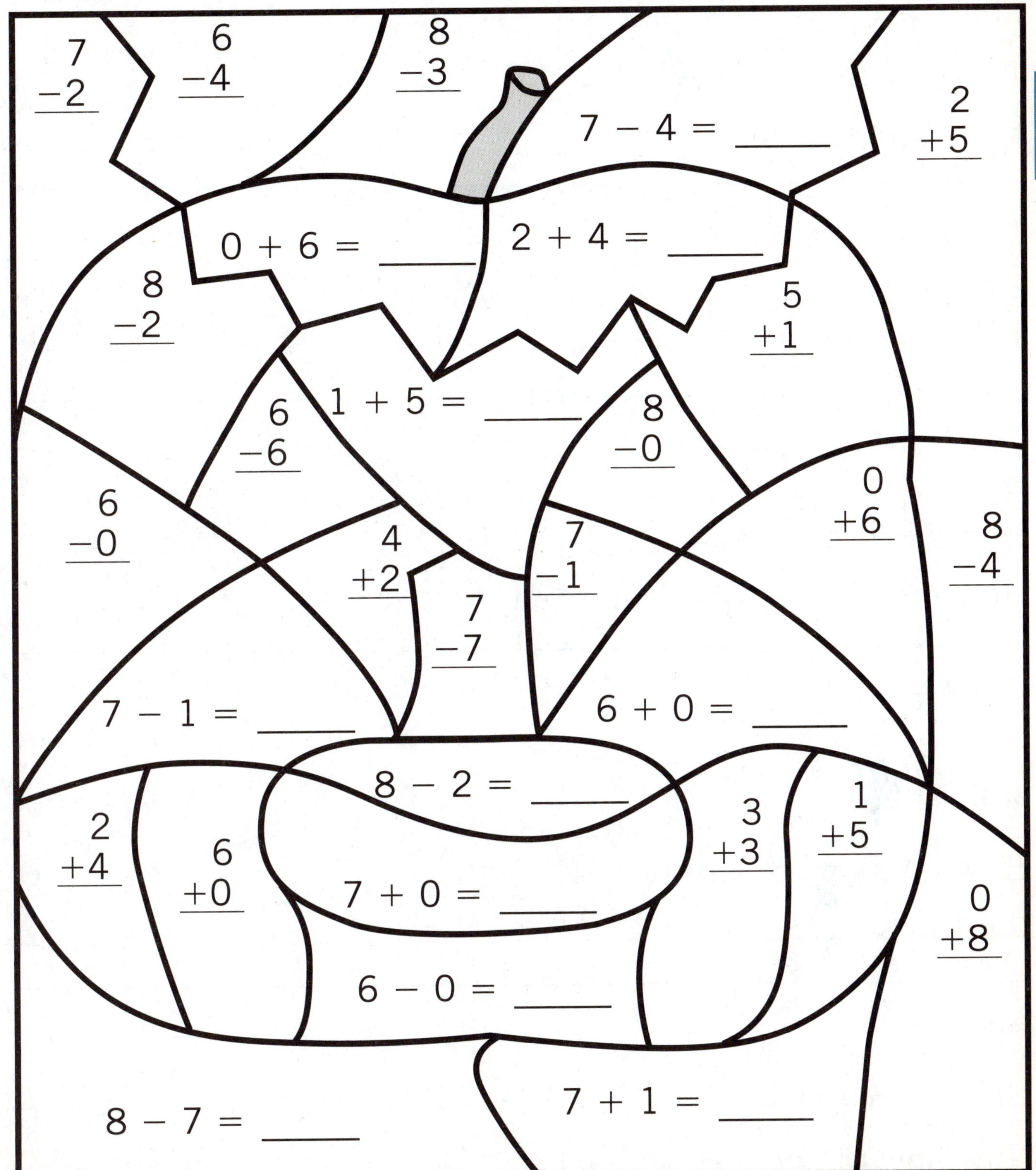

Lesson 10 Problem Solving

Solve each problem.

There are 5 white  .

There are 3 blue  .

How many in all?

There were 6  .

2 more came.

Then how many were there?

Joni needs 8  .

She has 5  .

How many more does she need?

---

7 white 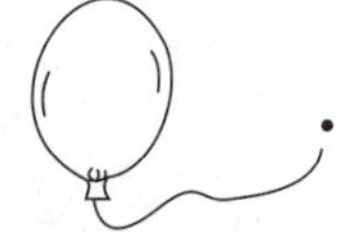 .

3 blue  .

How many more white 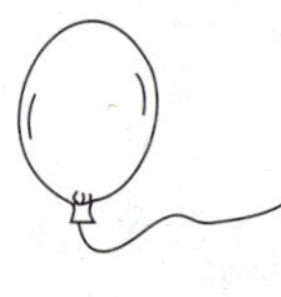

than blue  ?

Alex has 4  .

He finds 2 more.

Now how many does he have?

NAME ____________________

CHAPTER 3 PRACTICE TEST

Addition and Subtraction (facts through 8)

CHAPTER 3

Add.

3 +5	0 +6	4 +3	2 +5	2 +6
5 +1	4 +4	4 +2	1 +6	7 +1

Subtract.

6 −4	7 −2	8 −6	6 −3	8 −0
8 −1	7 −5	6 −2	6 −0	8 −3

Add or subtract. STOP! Watch the + and −.

3 +3	6 −1	7 −4	3 +4	8 −7
7 −7	5 +2	8 −4	7 +0	6 +2

NAME ______________________

CHAPTER 4 PRETEST
Addition and Subtraction (facts through 10)

CHAPTER 4

Add.

2 + 6	1 + 8	3 + 7	8 + 2	3 + 6
0 + 8	2 + 7	6 + 4	9 + 1	0 + 9
5 + 5	4 + 5	7 + 3	6 + 4	2 + 8

Subtract.

10 − 3	9 − 4	10 − 4	9 − 0	9 − 1
10 − 6	10 − 5	9 − 3	9 − 5	9 − 7
9 − 0	9 − 6	9 − 2	9 − 6	10 − 1

NAME ______________________

Lesson 1 Sums of 9

Add.

$\begin{array}{r} 2 \\ +7 \\ \hline 9 \end{array}$

2 + 7 = 9

$\begin{array}{r} 7 \\ +2 \\ \hline \end{array}$

7 + 2 = ____

$\begin{array}{r} 5 \\ +4 \\ \hline \end{array}$

5 + 4 = ____

$\begin{array}{r} 4 \\ +5 \\ \hline \end{array}$

4 + 5 = ____

$\begin{array}{r} 1 \\ +8 \\ \hline \end{array}$

1 + 8 = ____

$\begin{array}{r} 8 \\ +1 \\ \hline \end{array}$

8 + 1 = ____

$\begin{array}{r} 3 \\ +6 \\ \hline \end{array}$

3 + 6 = ____

$\begin{array}{r} 6 \\ +3 \\ \hline \end{array}$

6 + 3 = ____

$\begin{array}{r} 0 \\ +9 \\ \hline \end{array}$

0 + 9 = ____

$\begin{array}{r} 9 \\ +0 \\ \hline \end{array}$

9 + 0 = ____

$\begin{array}{r} 5 \\ +4 \\ \hline \end{array}$ $\begin{array}{r} 3 \\ +6 \\ \hline \end{array}$ $\begin{array}{r} 8 \\ +1 \\ \hline \end{array}$ $\begin{array}{r} 4 \\ +5 \\ \hline \end{array}$ $\begin{array}{r} 7 \\ +2 \\ \hline \end{array}$ $\begin{array}{r} 0 \\ +9 \\ \hline \end{array}$

CHAPTER 4

Lesson 2 Subtracting from 9

Subtract.

$\begin{array}{r} 9 \\ -6 \\ \hline 3 \end{array}$

9 − 6 = 3

$\begin{array}{r} 9 \\ -3 \\ \hline \end{array}$

9 − 3 = ____

$\begin{array}{r} 9 \\ -0 \\ \hline \end{array}$

9 − 0 = ____

$\begin{array}{r} 9 \\ -9 \\ \hline \end{array}$

9 − 9 = ____

$\begin{array}{r} 9 \\ -5 \\ \hline \end{array}$

9 − 5 = ____

$\begin{array}{r} 9 \\ -4 \\ \hline \end{array}$

9 − 4 = ____

$\begin{array}{r} 9 \\ -8 \\ \hline \end{array}$

9 − 8 = ____

$\begin{array}{r} 9 \\ -1 \\ \hline \end{array}$

9 − 1 = ____

$\begin{array}{r} 9 \\ -2 \\ \hline \end{array}$

9 − 2 = ____

$\begin{array}{r} 9 \\ -7 \\ \hline \end{array}$

9 − 7 = ____

Lesson 3 Sums of 10

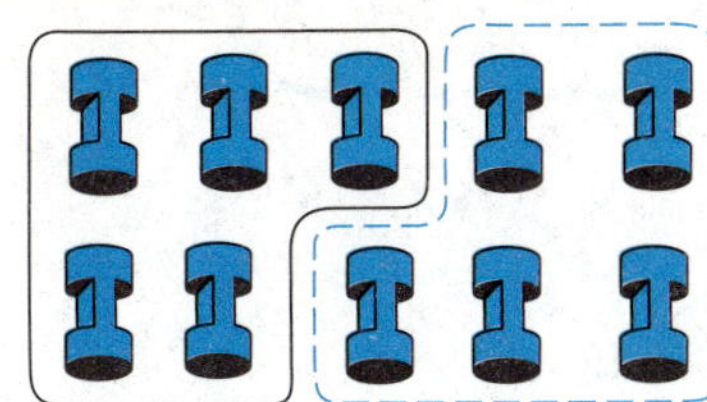

$5 + 5 = \underline{10}$

$\begin{array}{r} 5 \\ +5 \\ \hline 10 \end{array}$

Add.

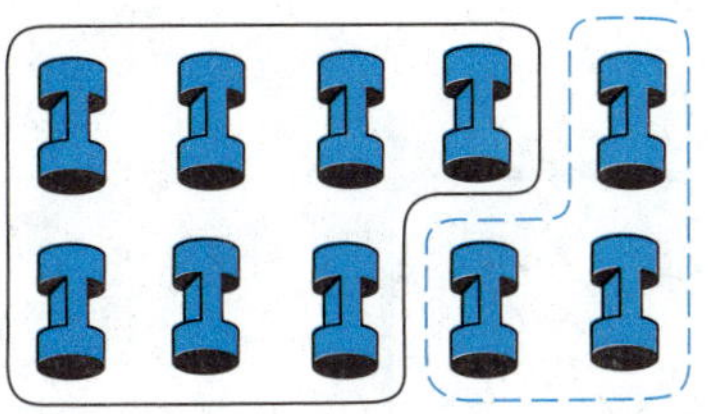

$7 + 3 = \underline{10}$

$\begin{array}{r} 7 \\ +3 \\ \hline 10 \end{array}$

$3 + 7 = \_\_\_\_$

$\begin{array}{r} 3 \\ +7 \\ \hline \end{array}$

$1 + 9 = \_\_\_\_$ $\begin{array}{r} 1 \\ +9 \\ \hline \end{array}$ $\begin{array}{r} 9 \\ +1 \\ \hline \end{array}$

$9 + 1 = \_\_\_\_$

$2 + 8 = \_\_\_\_$ $\begin{array}{r} 2 \\ +8 \\ \hline \end{array}$ $\begin{array}{r} 8 \\ +2 \\ \hline \end{array}$

$8 + 2 = \_\_\_\_$

$6 + 4 = \_\_\_\_$ $\begin{array}{r} 6 \\ +4 \\ \hline \end{array}$ $\begin{array}{r} 4 \\ +6 \\ \hline \end{array}$

$4 + 6 = \_\_\_\_$

$10 + 0 = \_\_\_\_$ $\begin{array}{r} 10 \\ +\ 0 \\ \hline \end{array}$ $\begin{array}{r} 0 \\ +10 \\ \hline \end{array}$

$0 + 10 = \_\_\_\_$

$\begin{array}{r} 3 \\ +7 \\ \hline \end{array}$ $\begin{array}{r} 6 \\ +4 \\ \hline \end{array}$ $\begin{array}{r} 5 \\ +5 \\ \hline \end{array}$ $\begin{array}{r} 1 \\ +9 \\ \hline \end{array}$ $\begin{array}{r} 2 \\ +8 \\ \hline \end{array}$ $\begin{array}{r} 7 \\ +3 \\ \hline \end{array}$

Lesson 4 Subtracting from 10

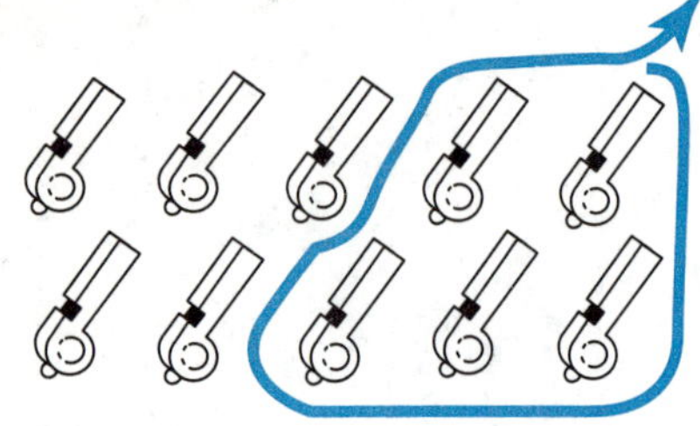

$\begin{array}{r} 10 \\ -5 \\ \hline 5 \end{array}$

10 − 5 = 5

$\begin{array}{r} 10 \\ -10 \\ \hline 0 \end{array}$

10 − 10 = 0

Subtract.

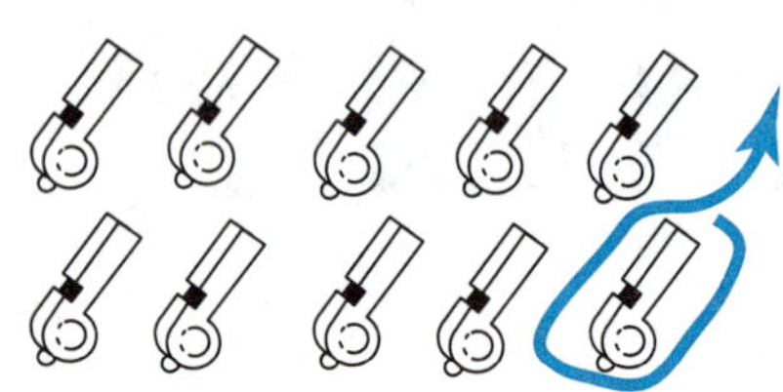

$\begin{array}{r} 10 \\ -1 \\ \hline \end{array}$

10 − 1 = ____

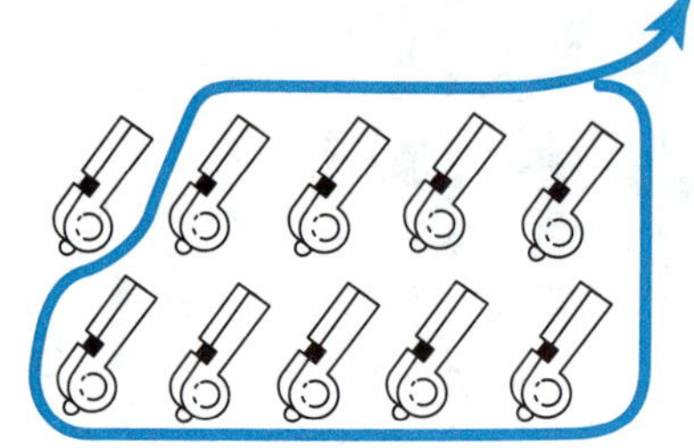

$\begin{array}{r} 10 \\ -9 \\ \hline \end{array}$

10 − 9 = ____

10 − 7 = ____

$\begin{array}{r} 10 \\ -7 \\ \hline \end{array}$ $\begin{array}{r} 10 \\ -3 \\ \hline \end{array}$

10 − 3 = ____

10 − 4 = ____

$\begin{array}{r} 10 \\ -4 \\ \hline \end{array}$ $\begin{array}{r} 10 \\ -6 \\ \hline \end{array}$

10 − 6 = ____

10 − 8 = ____

$\begin{array}{r} 10 \\ -8 \\ \hline \end{array}$ $\begin{array}{r} 10 \\ -2 \\ \hline \end{array}$

10 − 2 = ____

10 − 0 = ____

$\begin{array}{r} 10 \\ -0 \\ \hline \end{array}$

$\begin{array}{r} 10 \\ -7 \\ \hline \end{array}$ $\begin{array}{r} 10 \\ -1 \\ \hline \end{array}$ $\begin{array}{r} 10 \\ -5 \\ \hline \end{array}$ $\begin{array}{r} 10 \\ -10 \\ \hline \end{array}$ $\begin{array}{r} 10 \\ -2 \\ \hline \end{array}$ $\begin{array}{r} 10 \\ -6 \\ \hline \end{array}$

NAME ______________________________

Lesson 5 Practicing Addition

Add.

6	7	4	4	9	3
+4	+2	+4	+5	+1	+2
2	6	9	2	1	4
+7	+2	+0	+5	+4	+6
8	2	3	1	7	1
+1	+2	+6	+7	+3	+8
2	2	3	8	6	0
+3	+8	+5	+2	+1	+9
1	6	3	5	5	4
+9	+3	+4	+2	+4	+3
5	8	5	3	2	3
+3	+0	+5	+7	+6	+3

Solve each problem.

There are 5 white  .

There are 4 blue  .

How many in all?

There are 3 .

7 more  come.

How many are there now?

Beth has 9  .

She buys 1 more.

Now how many does she have?

There are 6 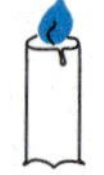 .

There are 3  .

How many in all?

There were 8  .

2 more came.

Then how many were there?

NAME ______________________

Lesson 6 Practicing Subtraction

Subtract.

9 −4	7 −6	10 −5	9 −7	8 −5	10 −9
10 −4	6 −3	9 −6	10 −3	9 −0	5 −1
3 −1	9 −1	10 −8	7 −2	9 −5	2 −2
10 −1	7 −0	5 −3	8 −7	10 −2	6 −4
9 −8	7 −4	10 −0	4 −2	8 −4	9 −3
10 −6	8 −6	9 −2	8 −1	9 −9	10 −7

CHAPTER 4

Lesson 6 Problem Solving

Solve each problem.

There are 10 white .

There are 4 blue .

How many more white than blue are there?

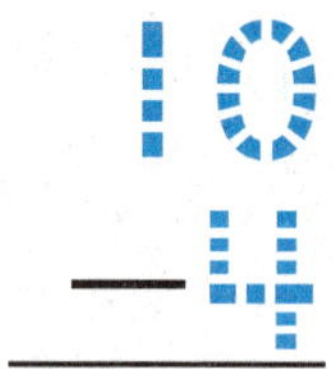

---

10 are on the table.

2 are broken.

How many are not broken?

There are 9  .

6 swim away.

How many  are left?

Joni wants 9  .

She has 5  .

How many more does she need?

There were 10 .

5 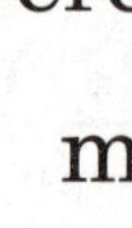 melted.

How many did not melt?

Lesson 7 Addition and Subtraction

Add or subtract. STOP! Watch the + and −.

5 + 2	10 − 6	7 + 1	9 − 4	1 + 9	9 − 8
9 + 0	8 − 0	10 − 9	1 + 8	2 + 7	9 − 1
10 − 1	2 + 8	1 + 6	8 − 3	6 + 3	10 − 3
9 − 9	4 + 5	8 + 1	10 − 8	9 − 0	5 + 5
4 + 4	9 − 3	7 + 3	9 − 7	3 + 6	10 − 7
7 + 2	8 − 7	0 + 8	9 − 2	10 − 5	4 + 6

Solve each problem.

There are 9 white .

There are 4 blue .

How many more white  than blue 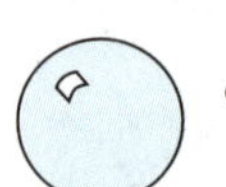 ?

Alex has 7 .

He finds 2 more.

Now how many does he have?

There are 5 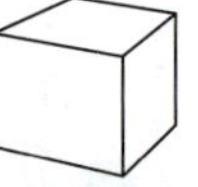.

There are 5 .

How many in all?

There were 10 .

6 ran away.

Then how many were left?

There were 9 .

8 were eaten.

How many were not eaten?

NAME ______________________

Lesson 8 Money

1 penny

1¢

1 nickel

5¢

1 dime

10¢

CHAPTER 4

Tell how much money.

______ ¢

______ ¢

______ ¢

______ ¢

______ ¢

______ ¢

______ ¢

______ ¢

Lesson 8 Problem Solving

10 pennies 10¢

1 dime

10¢

1 nickel

5¢

I have	I buy	I have left
	5¢	9¢ − 5¢ ¢
	6¢	¢ − ¢ ¢
	8¢	¢ − ¢ ¢
	5¢	¢ − ¢ ¢

Lesson 9 Problem Solving

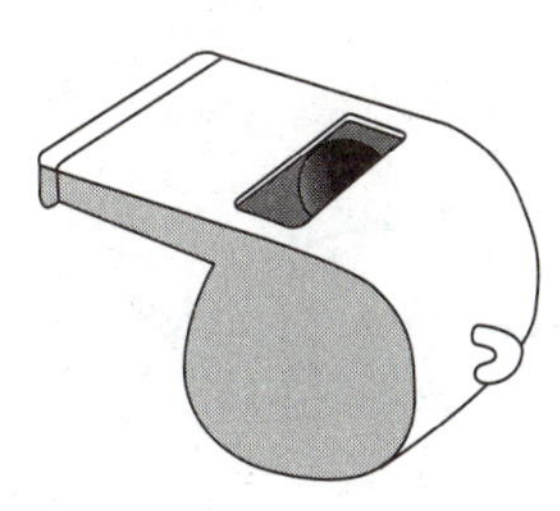

3¢ 4¢ 5¢ 6¢

CHAPTER 4

Add or subtract.

I buy  ¢
I buy  + 4¢
I spent ___¢

I buy
I buy  ___¢
+ ___¢
I spent ___¢

I have 1 0¢
I buy — ¢
I have left ___¢

I have 9¢
I buy — ___¢
I have left ___¢

I have 8¢
I buy — ___¢
I have left ___¢

I have 1 0¢
I buy — ___¢
I have left ___¢

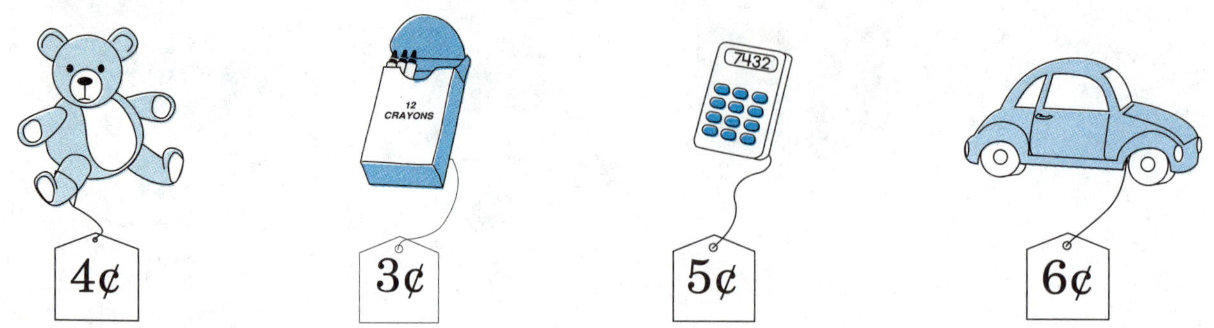

Add or subtract.

I buy ¢
I buy + ¢
I spent ¢

I buy ¢
I buy + ¢
I spent ¢

I have 8¢
I buy − ¢
I have left ¢

I have 1 0¢
I buy − ¢
I have left ¢

I have 7¢
I buy − ¢
I have left ¢

I buy ¢
I buy + ¢
I spent ¢

NAME ______________________

CHAPTER 4 PRACTICE TEST
Addition and Subtraction (facts through 10)

Add.

6	8	4	7	2
+2	+1	+6	+3	+8
5	6	1	3	9
+4	+3	+9	+6	+0

Subtract.

10	9	9	10	10
−4	−2	−6	−3	−1
9	9	10	9	10
−1	−5	−9	−7	−5

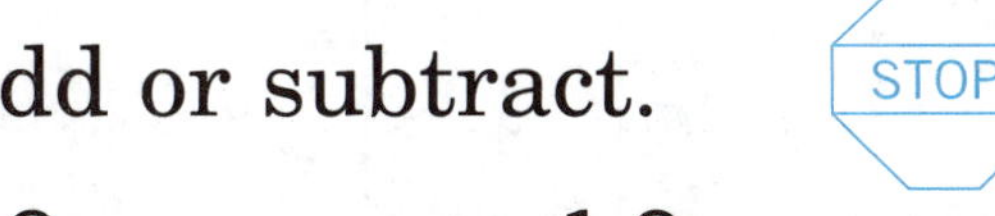

Add or subtract.

Watch the + and −.

0	10	9	2	9
+8	−8	−0	+7	−9
10	8	10	3	5
−2	+2	−6	+7	+5

CHAPTER 5 PRETEST
Numeration (0 through 99)

Complete.

1 ten	5 ones = ______	5 tens	0 ones = ______
2 tens	2 ones = ______	3 tens	4 ones = ______
6 tens	1 one = ______	4 tens	8 ones = ______
9 tens	6 ones = ______	3 tens	9 ones = ______
5 tens	5 ones = ______	8 tens	4 ones = ______
7 tens	4 ones = ______	6 tens	7 ones = ______
4 tens	2 ones = ______	5 tens	3 ones = ______

Write the numbers in order for each row.

12								20
32								40
60								68

NAME ______________________

Lesson 1 Numbers 10 through 19

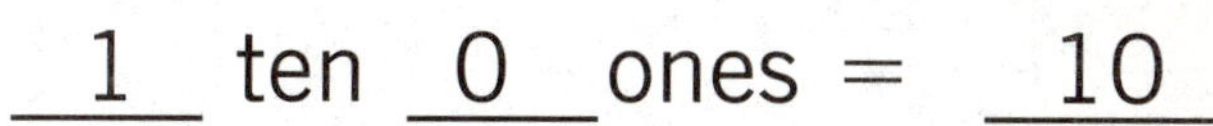

1 ten 0 ones = 10

1 ten 1 one = 11

Complete.

____ ten ____ ones = ____

____ ten ____ ones = ____

____ ten ____ ones = ____

____ ten ____ ones = ____

____ ten ____ ones = ____

____ ten ____ ones = ____

____ ten ____ ones = ____

____ ten ____ ones = ____

NAME ______________________

Lesson 2 Numbers 20 through 29

2 tens 0 ones = 20

2 tens 1 one = 21

Complete.

____ tens ____ ones = ____

____ tens ____ ones = ____

____ tens ____ ones = ____

____ tens ____ ones = ____

____ tens ____ ones = ____

____ tens ____ ones = ____

____ tens ____ ones = ____

____ tens ____ ones = ____

NAME ________________________

Lesson 3 Numbers 10 through 29

Complete.

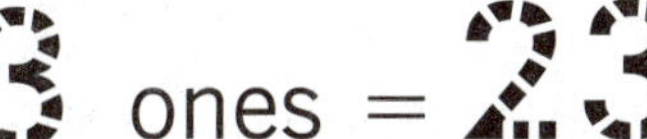
2 tens 3 ones = 23

____ tens ____ ones = ____

____ ten ____ ones = ____

____ tens ____ ones = ____

____ tens ____ ones = ____

____ ten ____ ones = ____

____ ten ____ ones = ____

____ tens ____ ones = ____

CHAPTER 5

NAME ____________________

Lesson 4 Tens

1 ten = 10

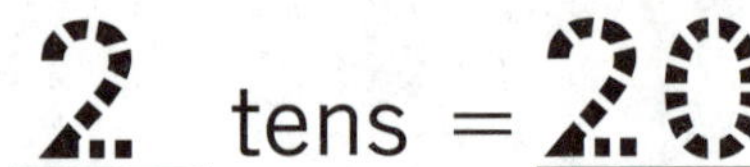

2 tens = 20

Complete.

____ tens = ____

____ tens = ____

____ tens = ____

____ tens = ____

____ tens = ____

____ tens = ____

____ tens = ____

NAME ______

Lesson 5 Numbers 30 through 49

Complete.

 tens  ones = 

__ tens ___ ones = ____

__ tens __ ones = ____

__ tens __ ones = ____

3 tens 7 ones = 37

4 tens 9 ones = ____

4 tens 3 ones = ____

3 tens 3 ones = ____

4 tens 6 ones = ____

4 tens 1 one = ____

3 tens 5 ones = ____

4 tens 2 ones = ____

4 tens 4 ones = ____

3 tens 8 ones = ____

CHAPTER 5

Complete the table.

0	1	2							9
10					15				
		22					27		
			33						
				44					49

Connect the dots in order.

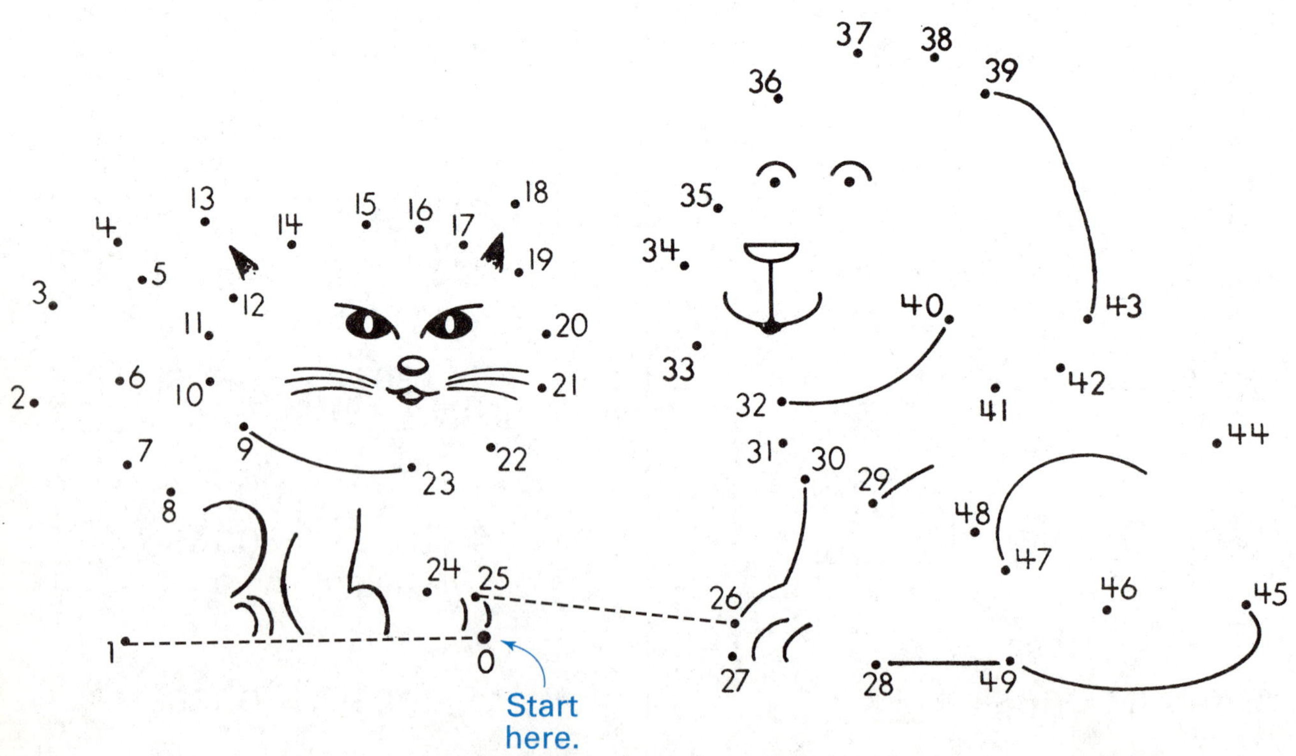

Lesson 6 Numbers 50 through 69

6 tens 4 ones = 64

Complete.

5 tens 0 ones = 50	6 tens 0 ones = ____
5 tens 1 one = ____	6 tens 1 one = ____
5 tens 2 ones = ____	6 tens 2 ones = ____
5 tens 3 ones = ____	6 tens 3 ones = ____
5 tens 4 ones = ____	6 tens 4 ones = ____
5 tens 5 ones = ____	6 tens 5 ones = ____
5 tens 6 ones = ____	6 tens 6 ones = ____
5 tens 7 ones = ____	6 tens 7 ones = ____
5 tens 8 ones = ____	6 tens 8 ones = ____
5 tens 9 ones = ____	6 tens 9 ones = ____

Lesson 7 Numbers 70 through 99

8 tens 6 ones = 86

9 tens 4 ones = 94

Complete.

7 tens 4 ones = ______

9 tens 1 one = ______

7 tens 8 ones = ______

9 tens 8 ones = ______

8 tens 5 ones = ______

9 tens 9 ones = ______

7 tens 0 ones = ______

8 tens 8 ones = ______

7 tens 9 ones = ______

8 tens 7 ones = ______

8 tens 9 ones = ______

9 tens 2 ones = ______

7 tens 3 ones = ______

9 tens 6 ones = ______

Lesson 8 Numbers 50 through 99

Tell how many cents.

58 ¢

____ ¢

____ ¢

____ ¢

____ ¢

____ ¢

____ ¢

____ ¢

Complete the table.

50		52						58	
60						66			69
							77		
80				84					
			93						99

Connect the dots in order.

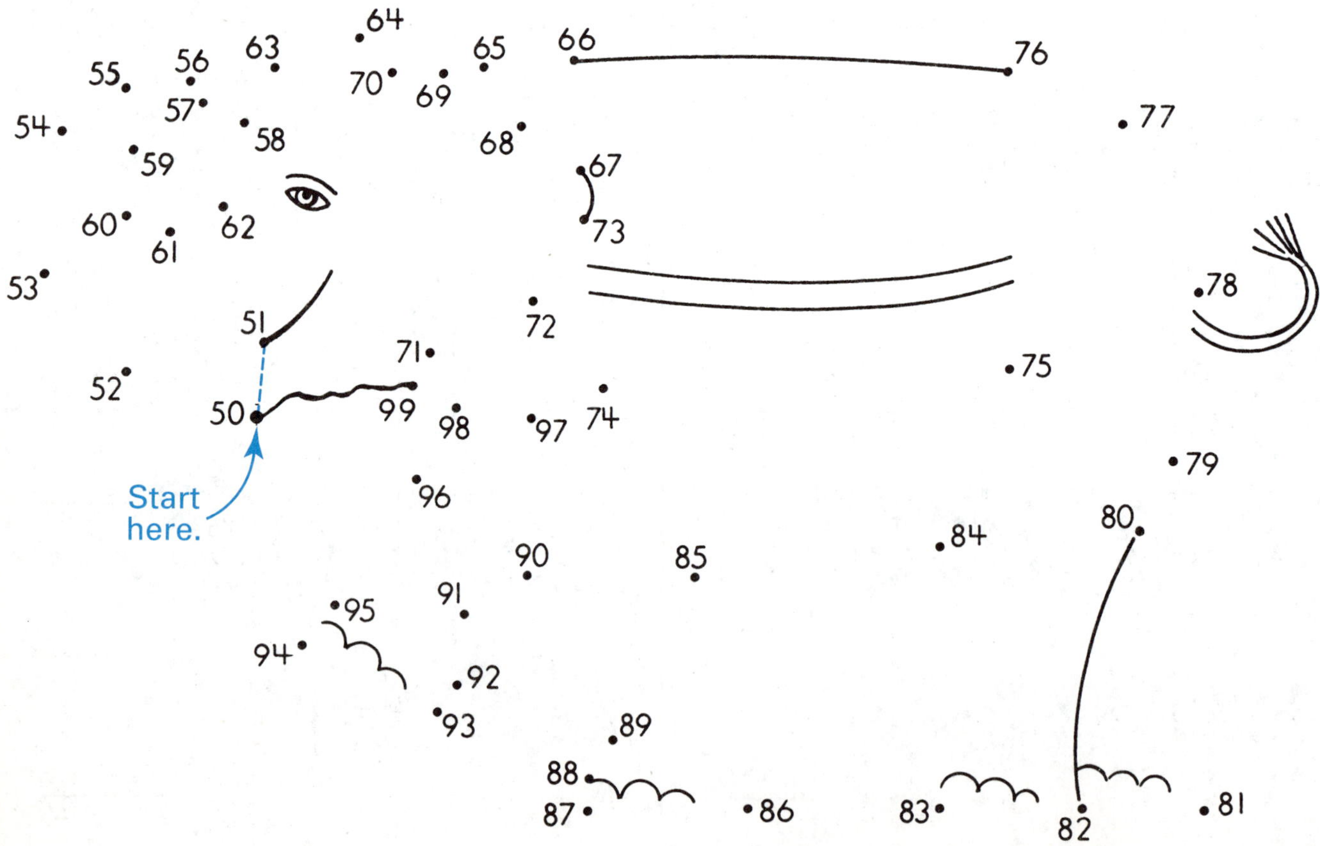

Lesson 9 Numeration

Ring the greatest and □ the least.

26 [15] (31)	53 71 68	40 70 90
32 23 30	85 65 95	64 46 42
36 92 67	91 57 19	78 94 29

Complete.

Before	*Between*	*After*
18, 19, 20	36, 37, 38	91, 92, ____
____, 64, 65	29, ____, 31	87, 88, ____
____, 70, 71	53, ____, 55	48, 49, ____
____, 34, 35	20, ____, 22	66, 67, ____

NAME ______________________

Lesson 10 Comparing Numbers

Ring the number that is greater.

10	22	37	25	28	38
25	28	49	43	33	37
95	97	74	76	58	50
20	30	21	23	42	47

Ring the number that is least.

35	24	73	53	24	17
22	29	46	40	53	52
68	64	92	98	36	35
43	49	25	31	39	51

NAME ____________________

Lesson 11 Skip Counting

Count by 2s. Write the missing numbers.

2, 4, 6, ____, 10, 12, 14

6, 8, 10, ____, 14, 16, 18

22, 24, ____, 28, 30, ____, 34

12, 14, ____, 18, 20, ____, 24

____, 32, 34, ____, 38, ____, 42

62, ____, 66, ____, 70, ____, 74

28, 30, ____, ____, 36, 38, 40, 42

50, ____, 54, 56, ____, ____, 62

84, ____, ____, 90, 92, ____, 96

46, 48, ____, 52, ____, ____, 58, 60

76, 78, 80, ____, ____, ____, 88

38, ____, ____, 44, ____, 48, 50

Count by 5s. Write the missing numbers.

5, 10, _____, 20, 25, 30

25, 30, _____, 40, 45, 50

60, 65, 70, 75, 80, _____, 90, _____

_____, 40, 45, _____, 55, 60, _____

50, _____, 60, 65, _____, _____, 80

15, 20, 25, _____, _____, _____, 45

Count by 10s. Write the missing numbers.

10, 20, _____, 40, 50, 60

20, 30, 40, _____, 60, _____, 80

50, _____, 70, 80, 90, _____

30, 40, _____, _____, 70, _____

0, 10, _____, 30, _____, _____

NAME ______________________

CHAPTER 5 PRACTICE TEST
Numeration (0 through 99)

Complete.

3 tens 7 ones = _____ 9 tens 9 ones = _____

2 tens 8 ones = _____ 3 tens 1 one = _____

7 tens 0 ones = _____ 5 tens 4 ones = _____

Name the numbers in order for each row.

16								24
79								87

Ring the number that is greater.

16 19 | 35 31 | 76 72

Count by 2s. Write the missing numbers.

4, 6, _____, 10, _____, 14

26, 28, 30, _____, 34, _____, 38

NAME ____________________

CHAPTER 6 PRETEST
Measurement

CHAPTER 6

Write the time for each clock.

6:30

_____ thirty

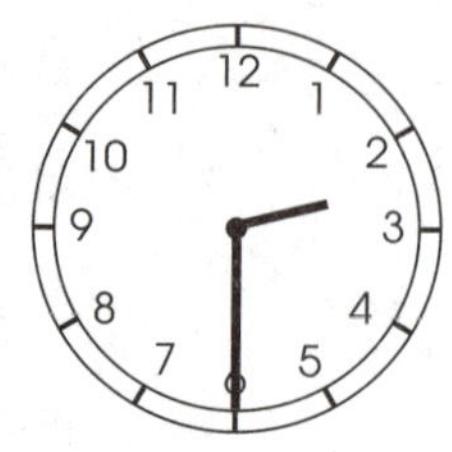

___ : _____

_____ : _____

Use a centimeter ruler.

How long is each object?

____ centimeters

____ centimeters

Use an inch ruler.

How long is each object?

____ inches

____ inches

NAME ______________________

Lesson 1 Time—Hour

4:00

4 o'clock
4:00

Both clocks show the same time.

CHAPTER 6

Write the time for each clock.

 o'clock	____ o'clock	____ o'clock
____ : ____	____ : 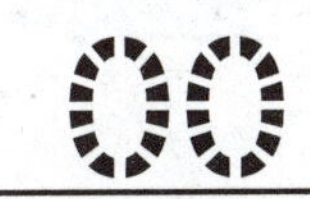	____ : 
	5:00	
____ o'clock	____ o'clock	____ o'clock
____ : ____	____ : ____	____ : ____

____ o'clock	____ o'clock	____ o'clock
____ : ____	____ : ____	____ : ____

Show this time on this clock. Show this time on this clock.

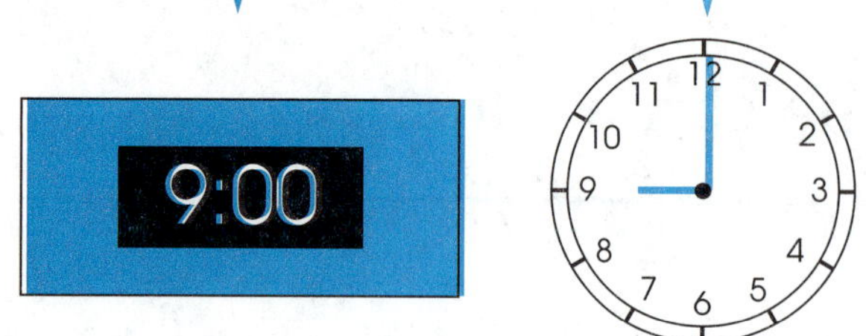

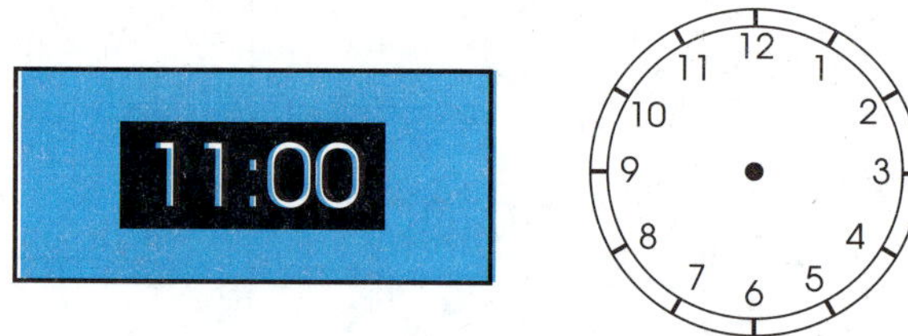

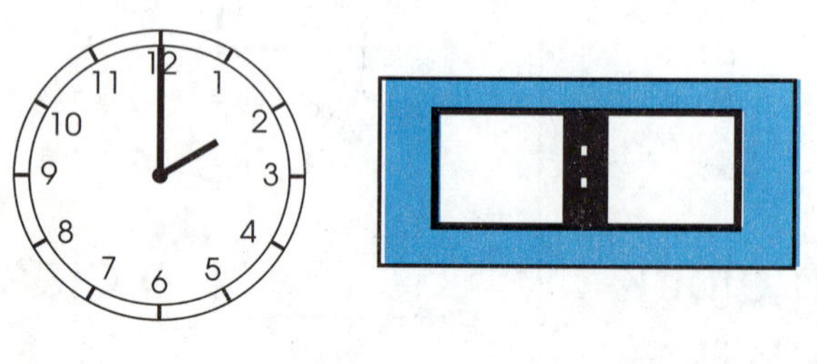

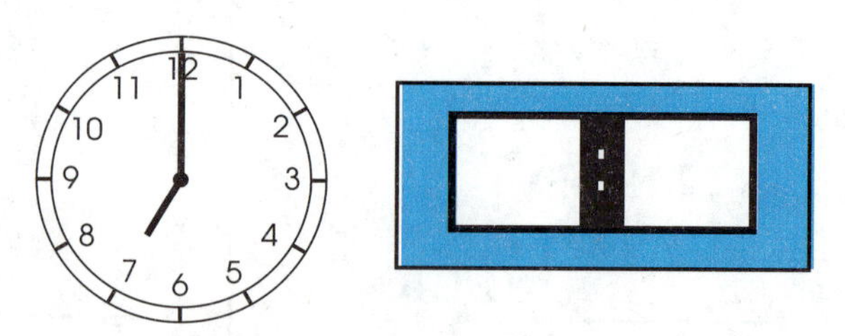

Lesson 2 Time—Half Hour

1 o'clock
1:00

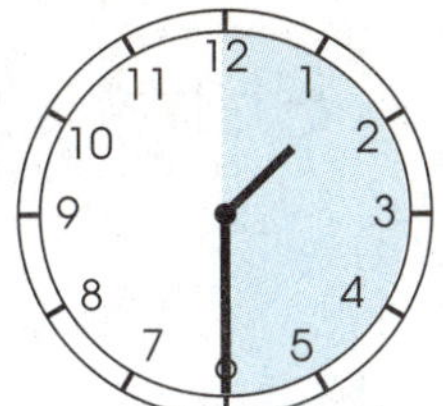

one thirty
1:30

2 o'clock
2:00

Write the time for each clock.

two thirty
2 : 30

_____ thirty

_____ : 30

_____ thirty
_____ : 30

_____ thirty
_____ : _____

_____ thirty
_____ : _____

_____ thirty

_____ : _____

_____ thirty
_____ : _____

_____ thirty
_____ : _____

_____ thirty
_____ : _____

CHAPTER 6

Lesson 2 Time—Half Hour

Show this time on this clock. Show this time on this clock.

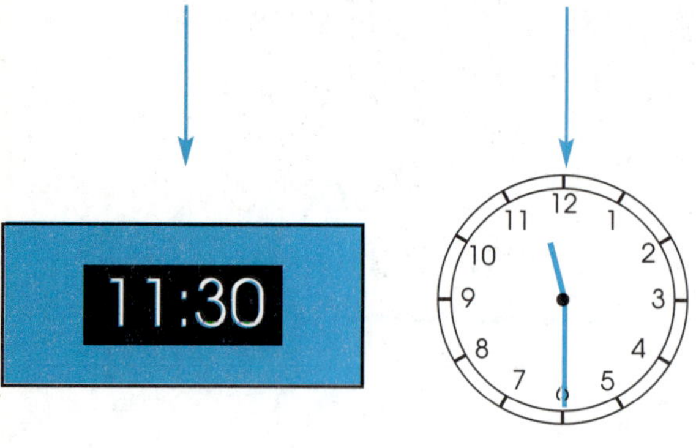

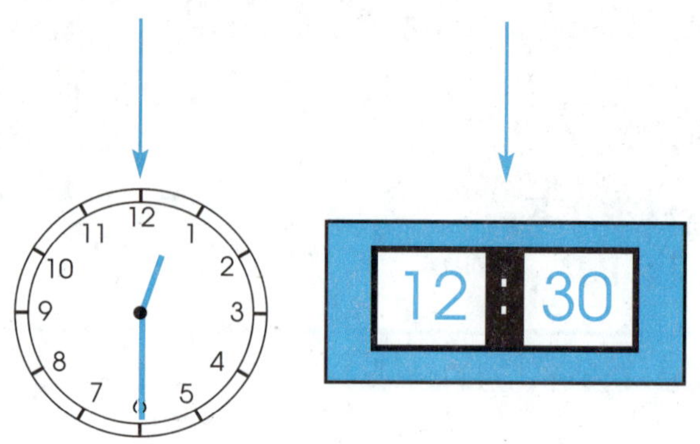

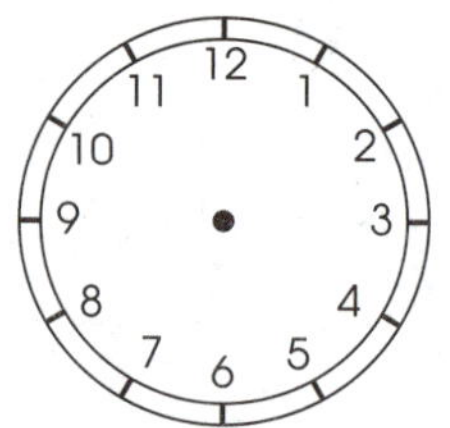

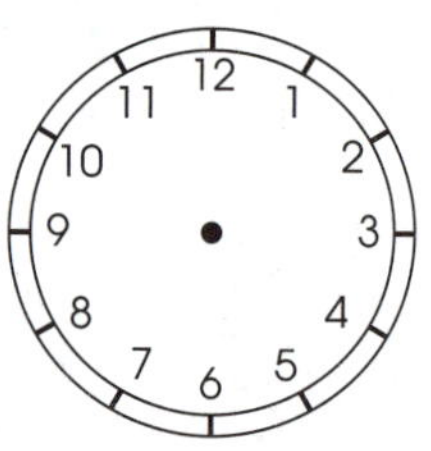

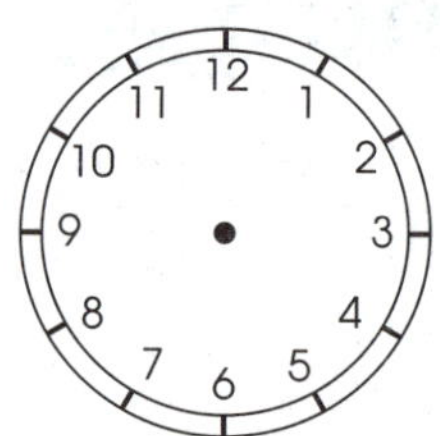

Lesson 3 Calendar

September						
S	M	T	W	Th	F	S
	1	2	3	4	5	6
7	8	9	10	11	12	13
14	15	16	17	18	19	20
21	22	23	24	25	26	27
28	29	30				

There are 12 months in a year.
September has 30 days.
September 1 is on Monday.
There are 5 Mondays in September.
There are 4 Saturdays in September.

September 19 is on __________.

Complete.

How many days are in a week? ____________

What day comes after Thursday? __________

September 30 is on ____________ .

There are ______ Tuesdays in September.

Days of the Week
Sunday
Monday
Tuesday
Wednesday
Thursday
Friday
Saturday

October						
S	M	T	W	Th	F	S
			1	2	3	4
5	6	7	8	9	10	11
12	13	14	15	16	17	18
19	20	21	22	23	24	25
26	27	28	29	30	31	

October has ______ days.

October 1 is on __________________.

There are ______ Wednesdays in October.

October 31 is on __________ .

There are ______ Sundays in October.

Lesson 3 Calendar

January						
Sun.	Mon.	Tues.	Wed.	Thurs.	Fri.	Sat.
				1	2	3
4	5	6	7	8	9	10
11	12	13	14	15	16	17
18	19	20	21	22	23	24
25	26	27	28	29	30	31

Months of the Year	*Number of Days*
January	31
February	28
March	31
April	30
May	31
June	30
July	31
August	31
September	30
October	31
November	30
December	31

Complete.

Thirty days have September,
April, June, and ________________.
All the rest have thirty-one,
Except ________________, which has only 28.

The first month of the year is ______________.

The last month of the year is ________________.

January has ______ days.

January 3 is on ______________.

In what month is your birthday? ____________________

What month is it today? ____________________

In what month is Valentine's Day? ______________

NAME ____________________

Lesson 4 Charts

March		Our Weather Calendar				
Sun.	Mon.	Tues.	Wed.	Thurs.	Fri.	Sat.
				1	2	3
4	5	6	7	8	9	10
11	12	13	14	15	16	17
18	19	20	21	22	23	24
25	26	27	28	29	30	31

sunny

rainy

cloudy

snowy

stormy

Complete.

What was the weather on

March 4? ____________ March 10? ____________

March 15? ____________ March 21? ____________

March 30? ____________ March 31? ____________

How many days did it rain  ? ______

How many days did it storm ? ______

How many days did it snow  ? ______

How many days was it cloudy   ? ______

How many days was it sunny  ? ______

CHAPTER 6

Make a weather calendar for one week.

Write the numbers to show the dates.

Draw to show the weather.

sunny

cloudy

rainy

snowy

stormy

Sunday	Monday	Tuesday	Wednesday	Thursday	Friday	Saturday

Use your weather calendar.

How many days was it

cloudy  ? ______

sunny  ? ______

snowy  ? ______

stormy ? ______

rainy  ? ______

Complete.

Write today's date. ______ ______, ______
Month Date Year

What day of the week is it today? ______

When were you born? ______ ______, ______
Month Date Year

Lesson 5 Bar Graphs

Look at the bar graph.

Animals

10
9
8
7
6
5
4
3
2
1

How many cats? 

How many fish? ☐

How many dogs? 

How many rabbits? ☐

Count each kind.

Color squares to show how many.

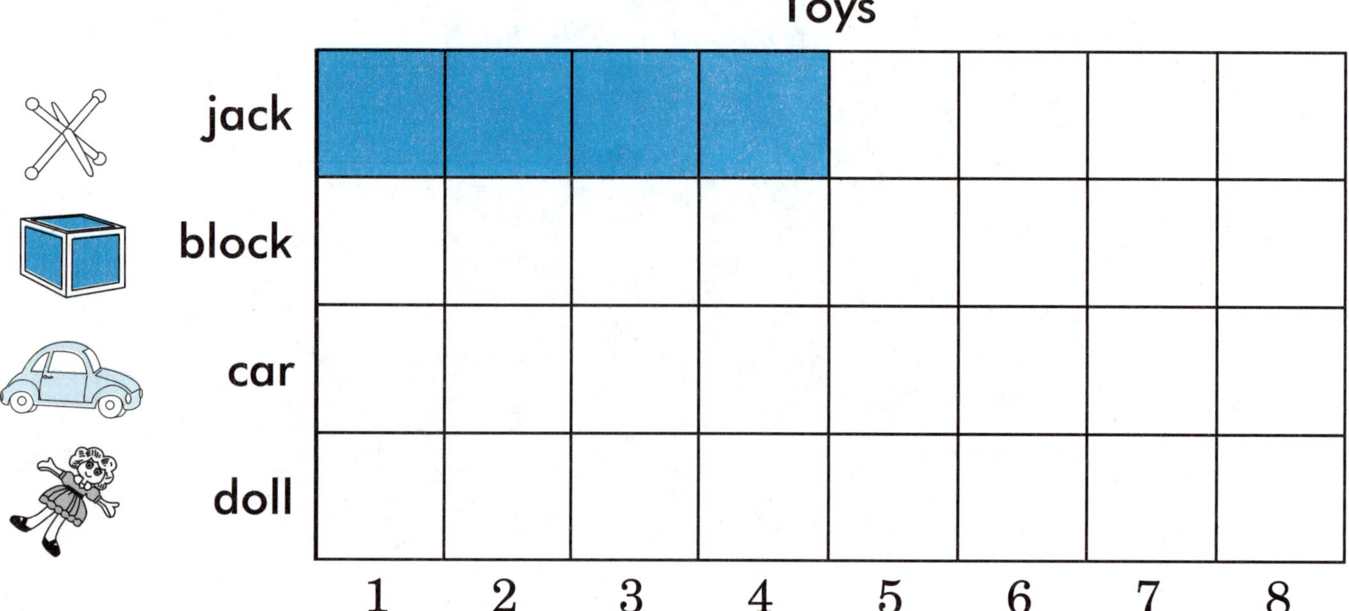

How many jacks? ☐

How many cars? ☐

How many blocks? ☐

How many dolls? ☐

Lesson 6 Picture Graphs

Look at the picture graph.

Our Favorite Lunches

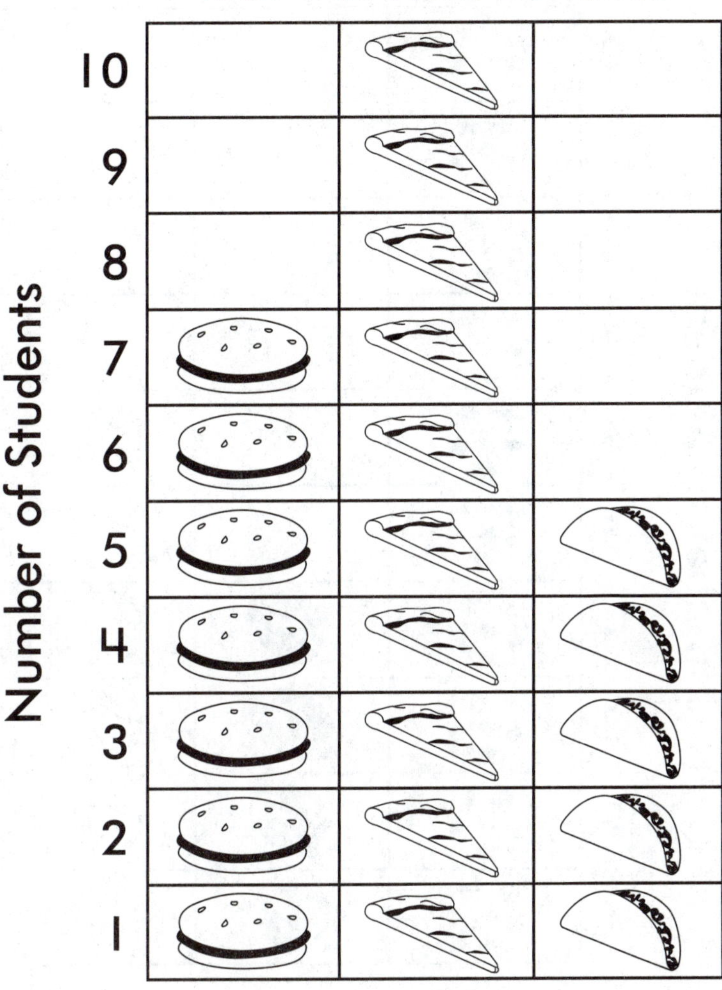

How many students like  ? 

How many students like 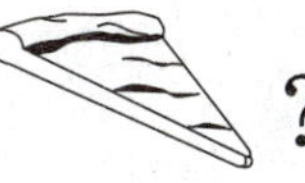 ? 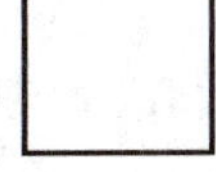

How many students like  ? 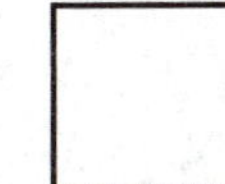

Lesson 6 Picture Graphs

Look at the picture graph.

Our Favorite Sport

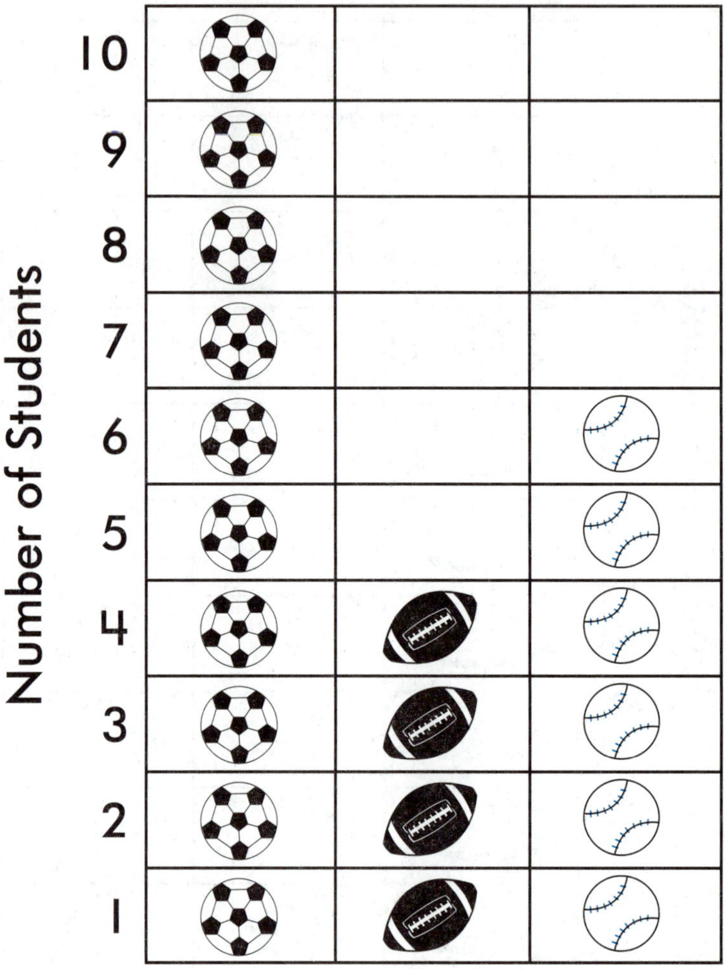

Number of Students

10			
9			
8			
7			
6			
5			
4			
3			
2			
1			

How many students like  ? ☐

How many students like  ? ☐

How many students like  ? ☐

Lesson 7 Centimeter

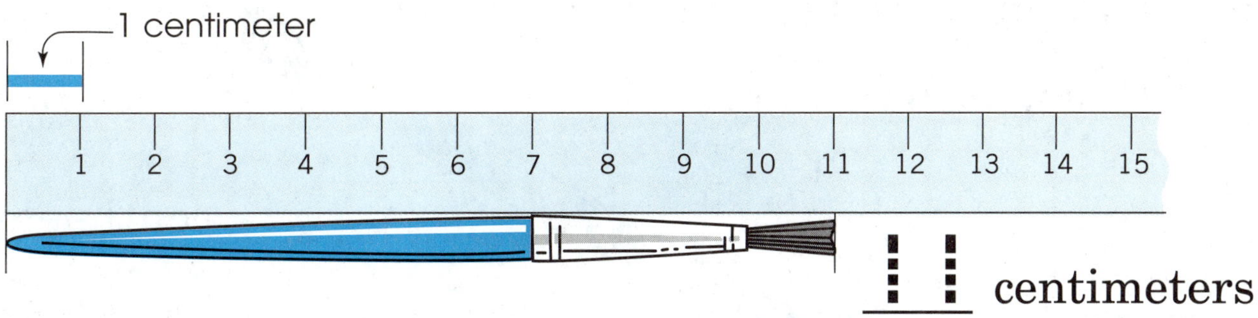

11 centimeters

How long is each object?

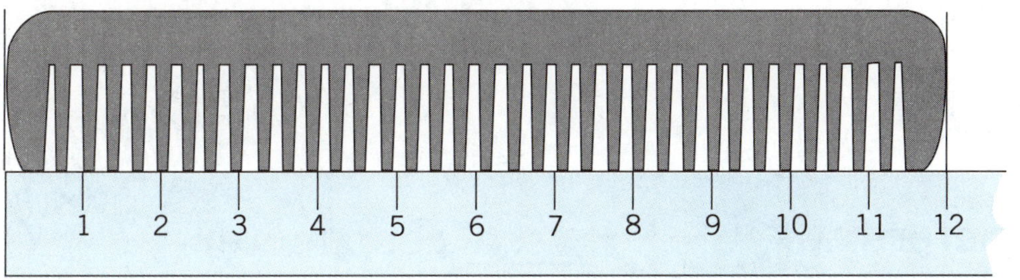

____ centimeters

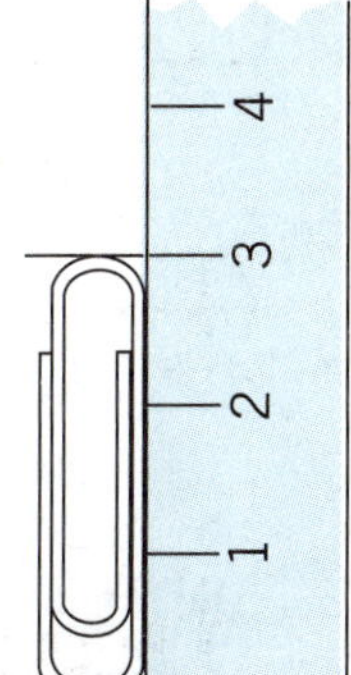

____ centimeters

____ centimeters

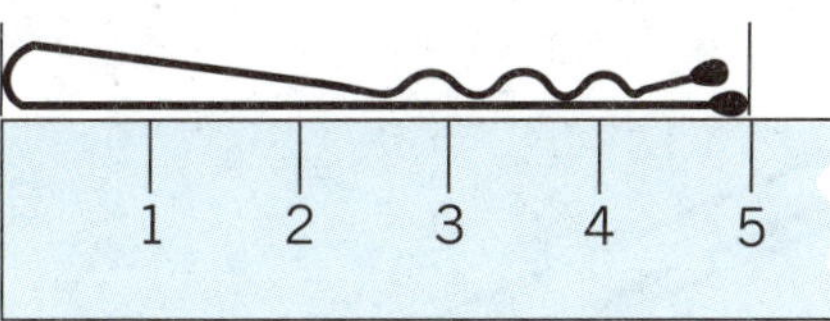

____ centimeters

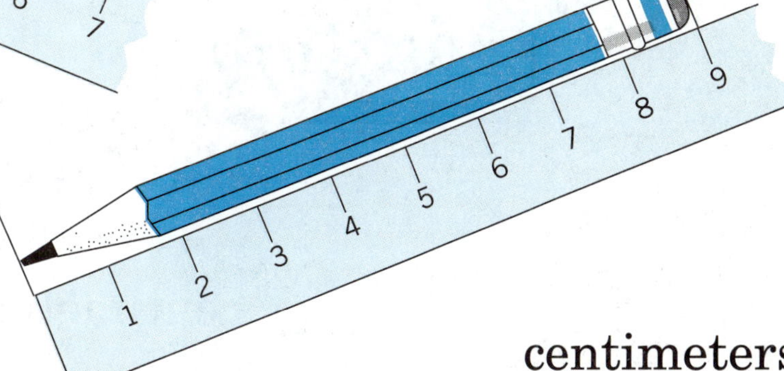

____ centimeters

Lesson 7 Centimeter

←Cut off and save this ruler.

How long is each object?

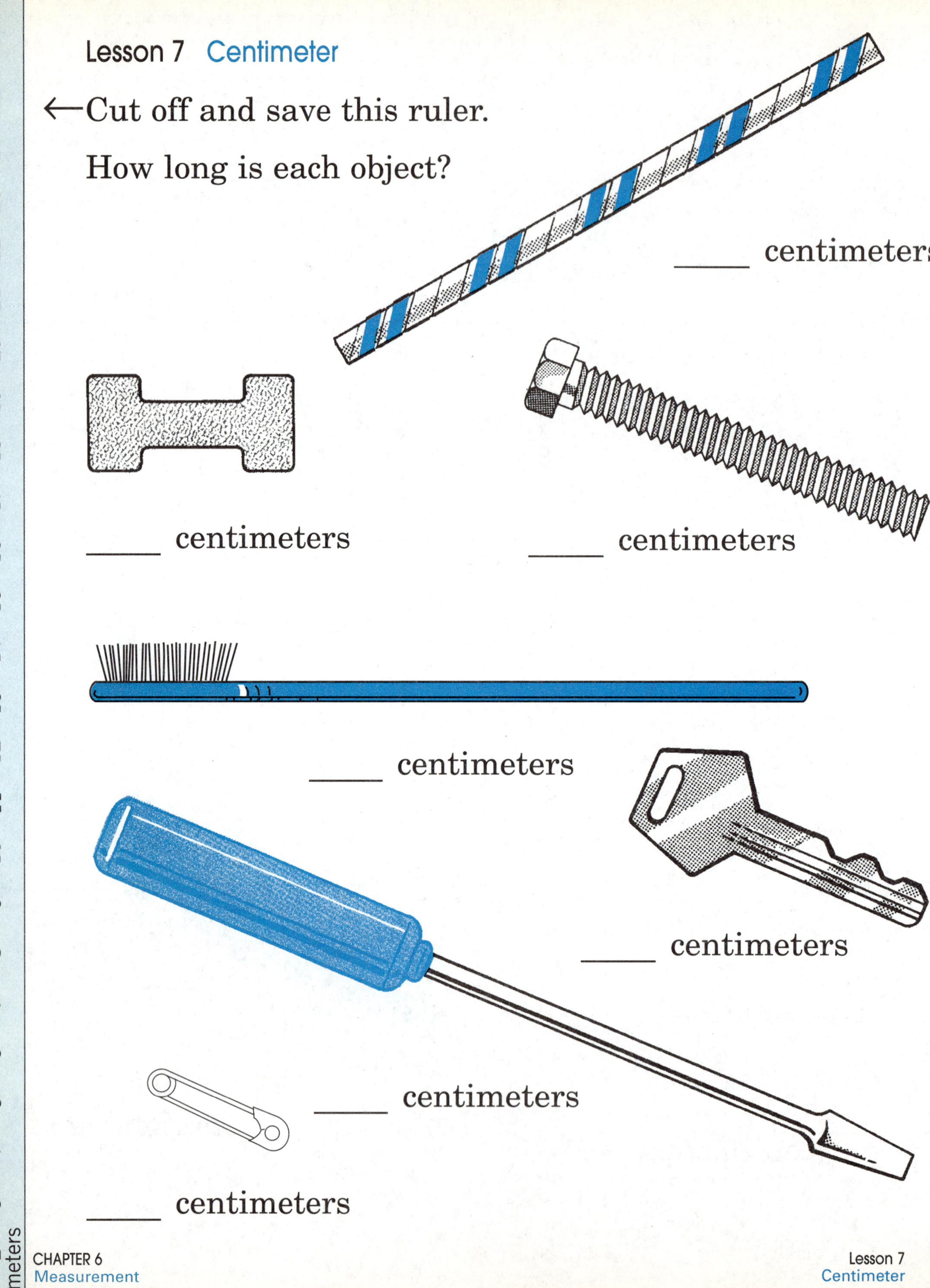

____ centimeters

____ centimeters

____ centimeters

____ centimeters

____ centimeters

____ centimeters

____ centimeters

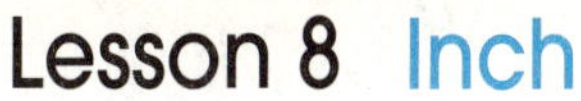

Lesson 8 Inch

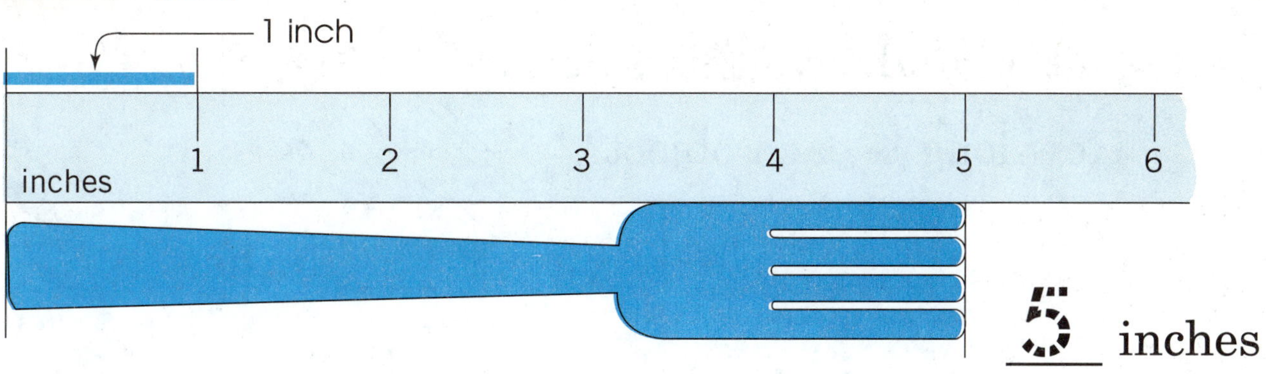

5 inches

How long is each object?

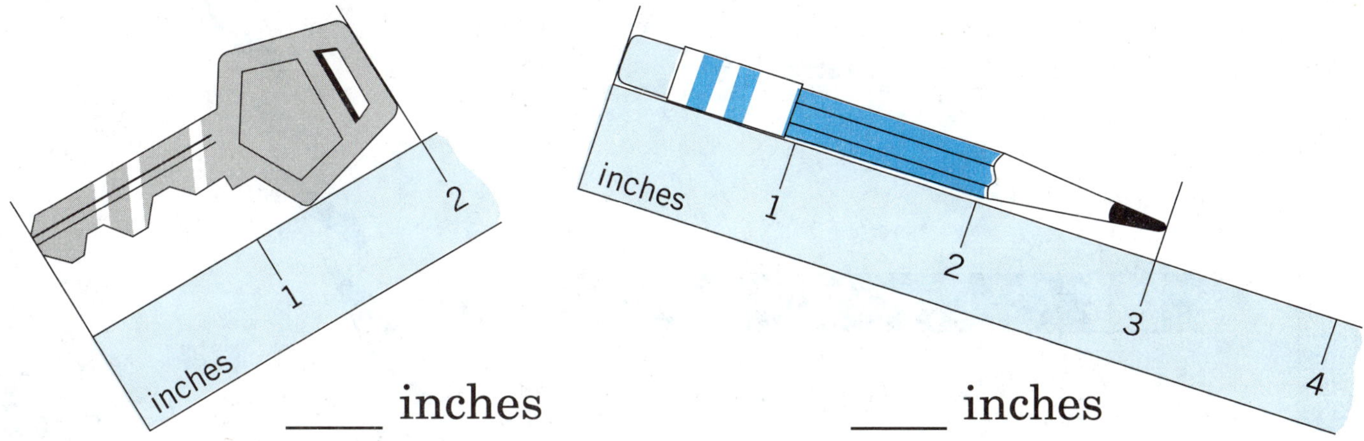

____ inches ____ inches

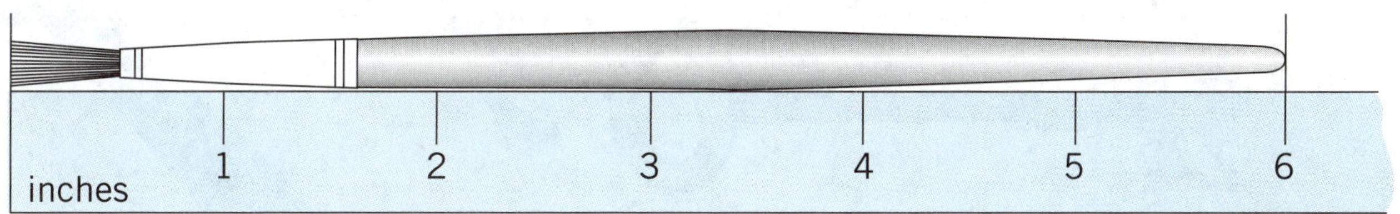

____ inches

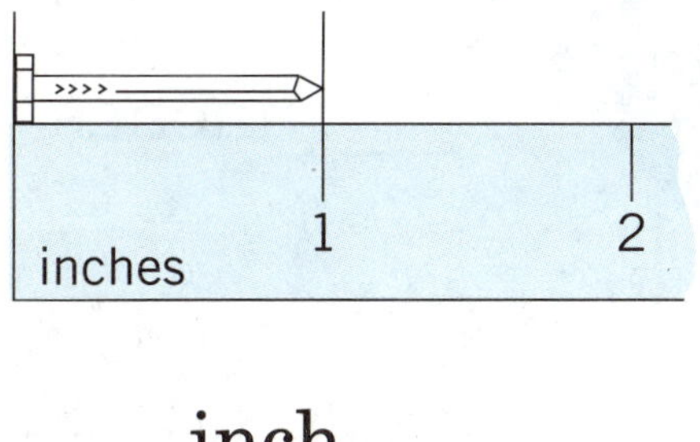

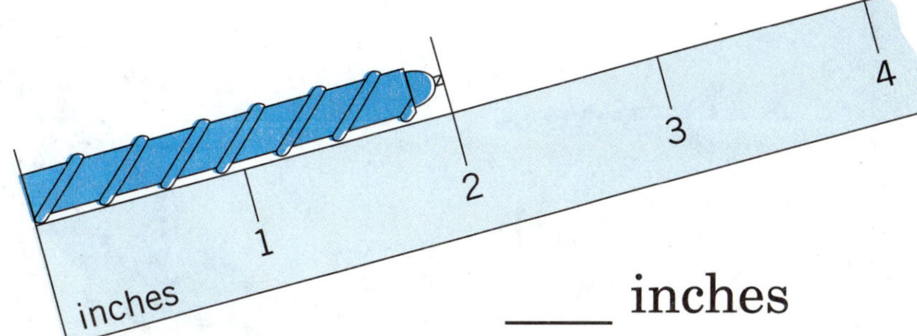

____ inch ____ inches

Lesson 8 Inch

←Cut off and save this ruler.

How long is each object?

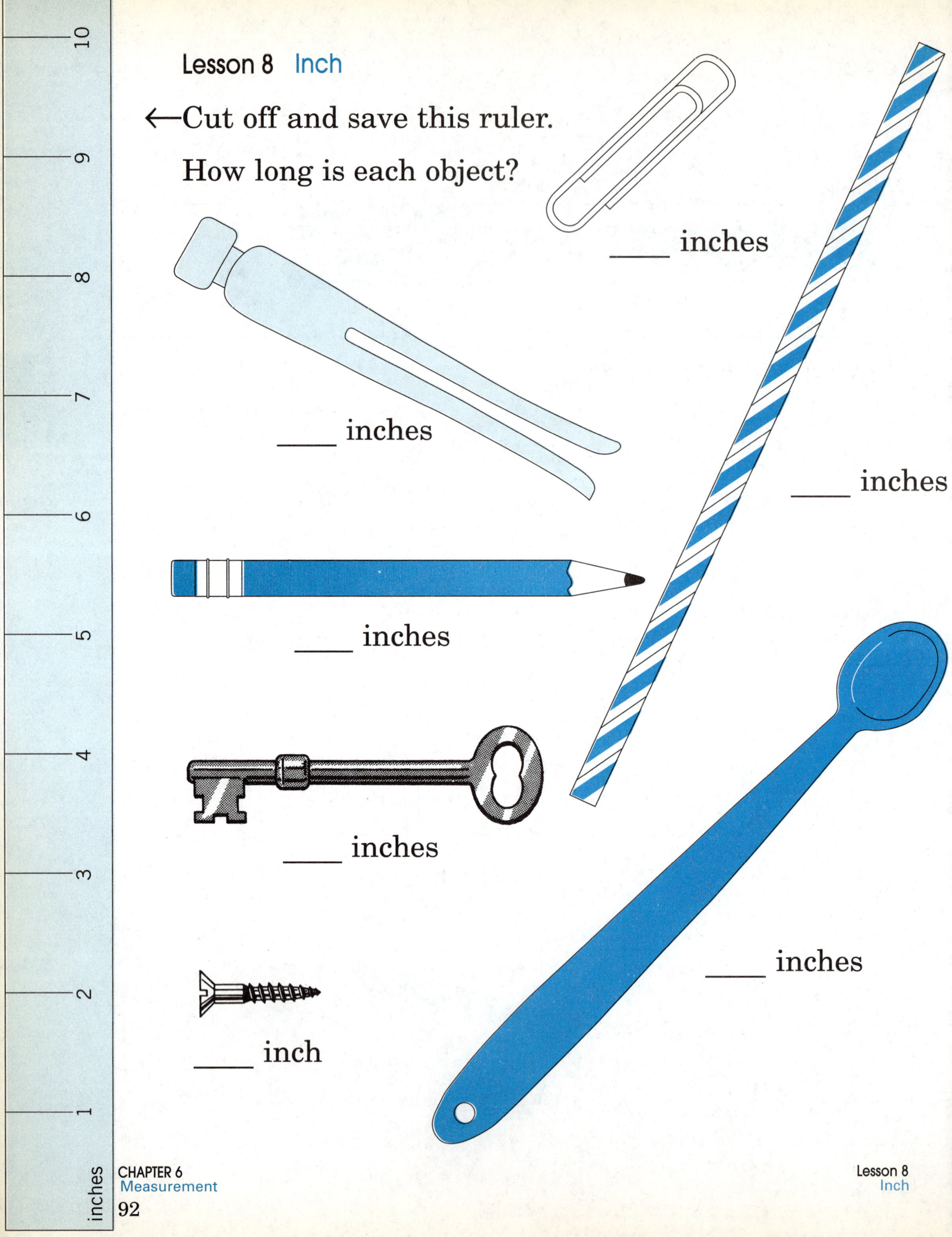

NAME ______________________

Lesson 9 Measuring

Work with a friend.
Use a centimeter ruler.
Measure each other.

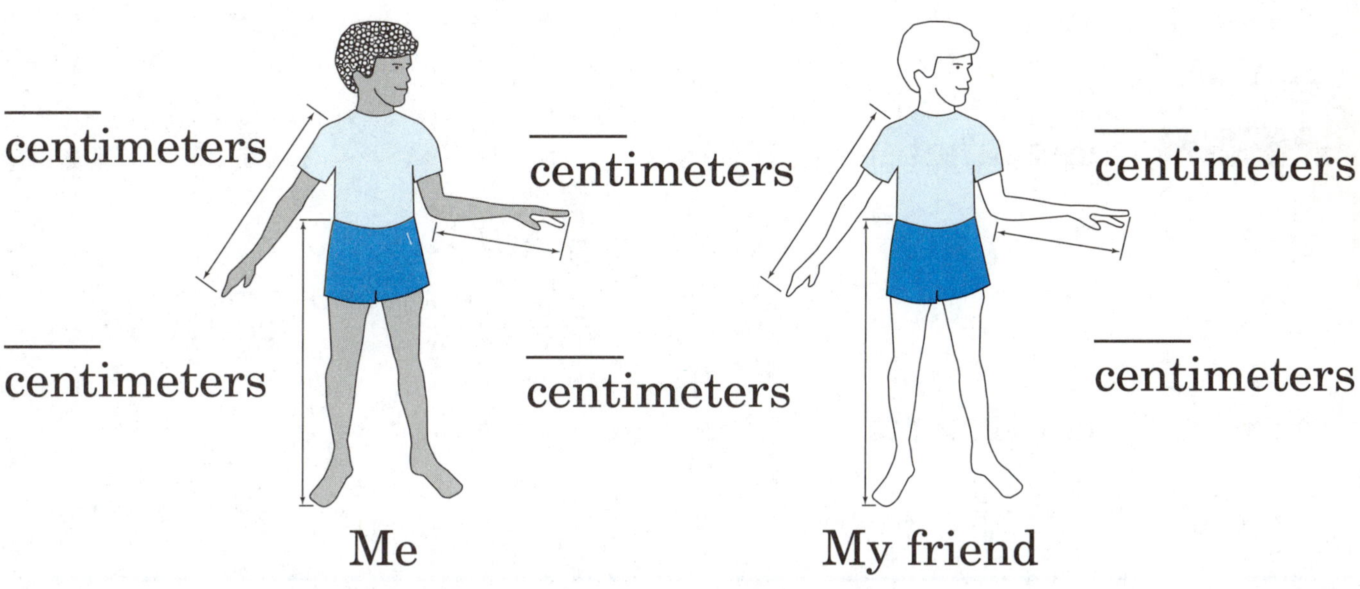

Use an inch ruler.
Measure each other.

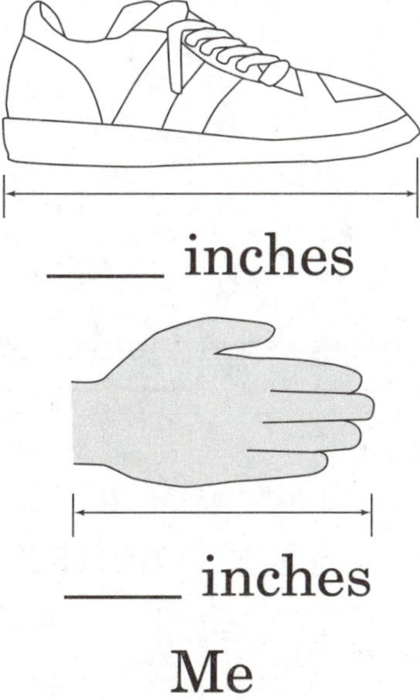

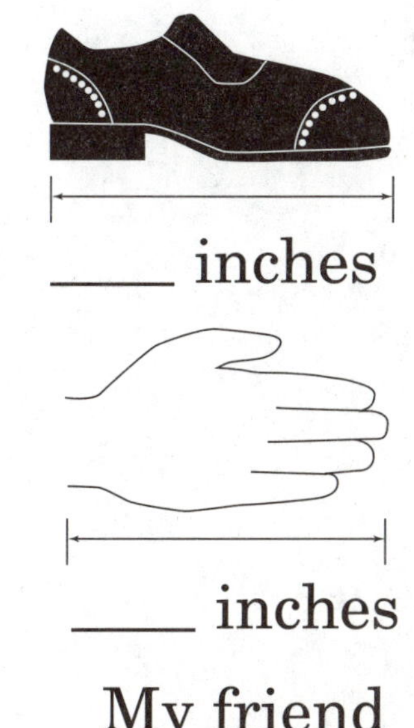

Lesson 9 Measuring

Work with a friend.

Use a centimeter ruler.

Measure these things found in your classroom.

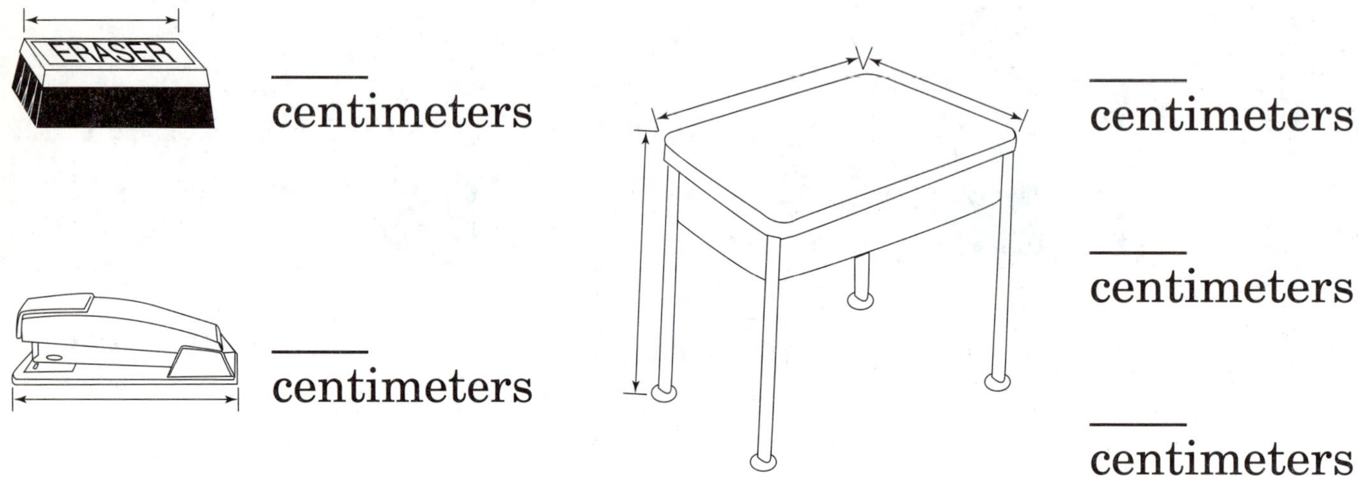

Use an inch ruler.

Measure these things found in your classroom.

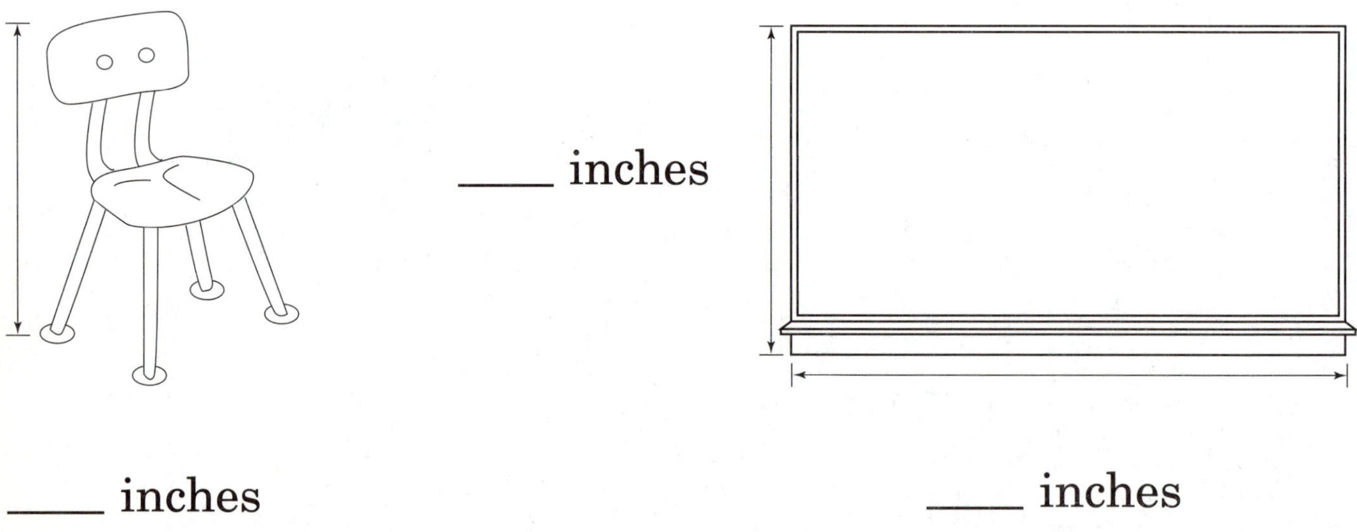

NAME ______________________

CHAPTER 6 PRACTICE TEST
Measurement

Write the time for each clock.

7:30

_______ thirty

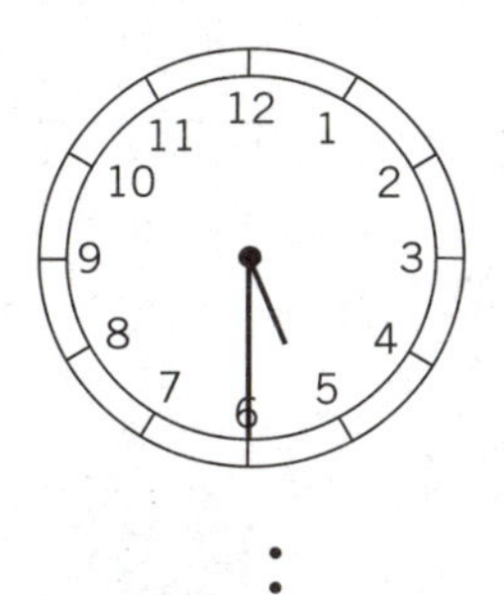

_____ : _____

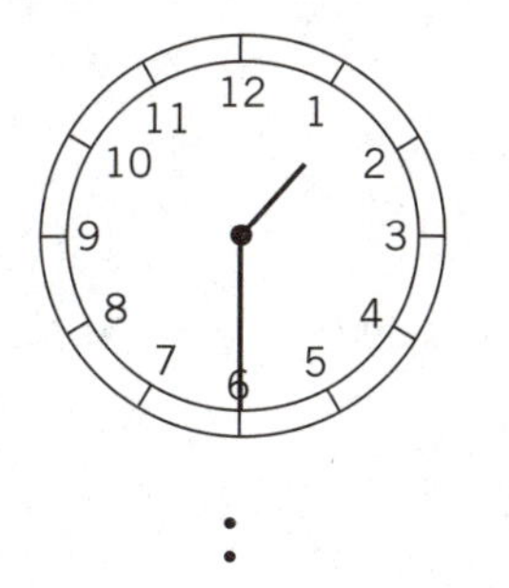

_____ : _____

Use a centimeter ruler.

How long is each object?

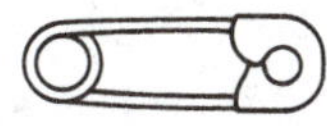

____ centimeters

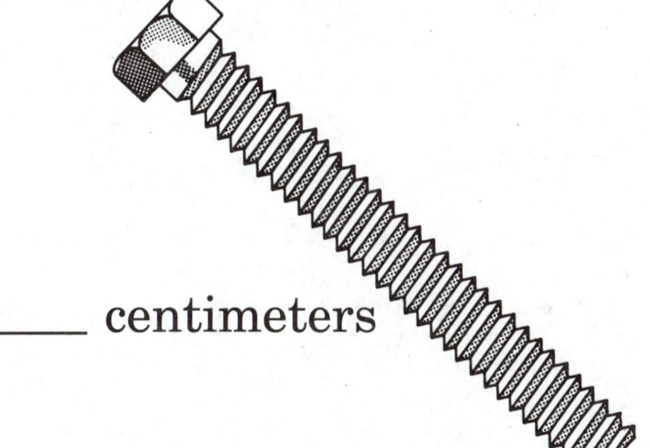

____ centimeters

Use an inch ruler.

How long is each object?

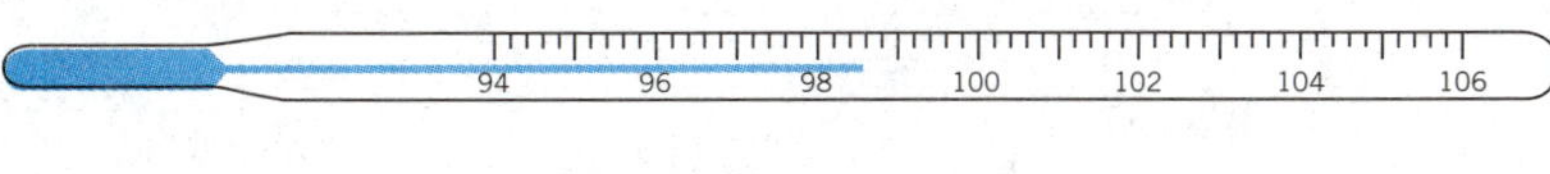

____ inches

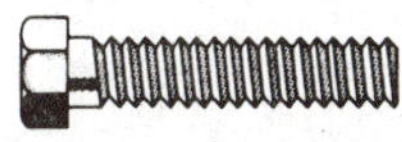

____ inch

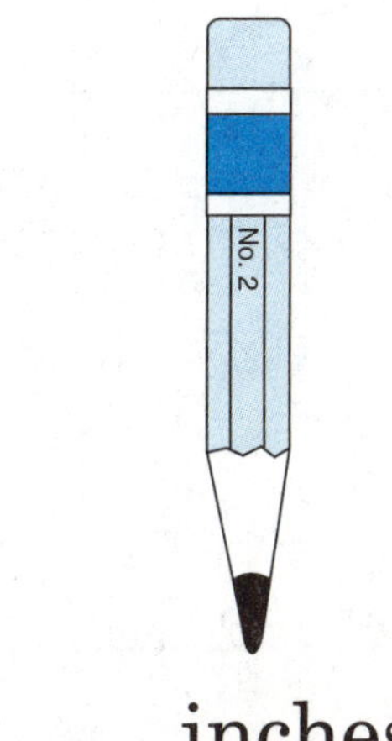

____ inches

CHAPTER 7 PRETEST
Geometry

Ring the shape.

square  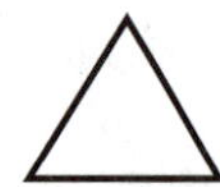 

cube  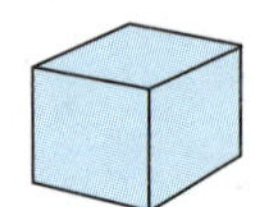 

circle 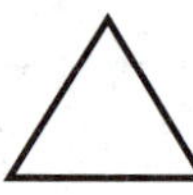  

sphere 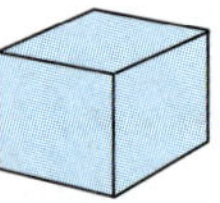  

Find the pattern. Ring the shape that comes next.

   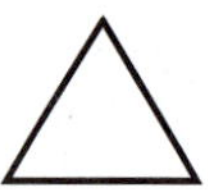  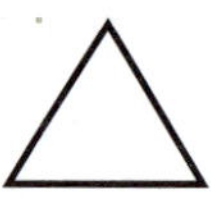  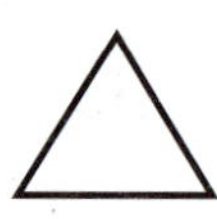 

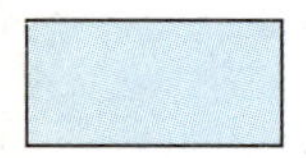 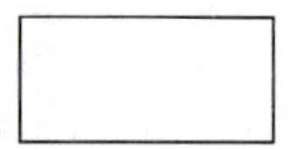    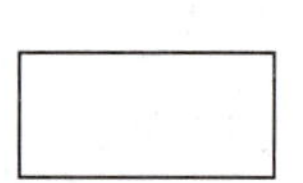 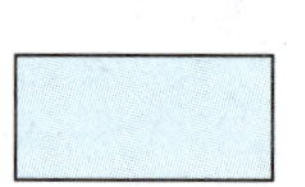 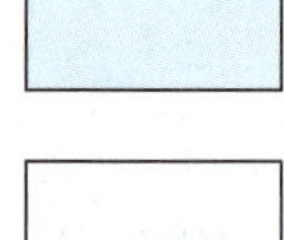

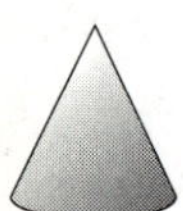    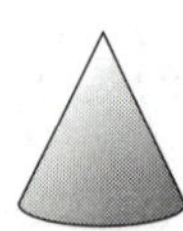   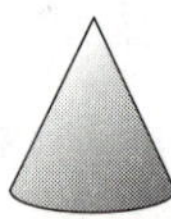

NAME ______________________________

Lesson 1 Plane Figures

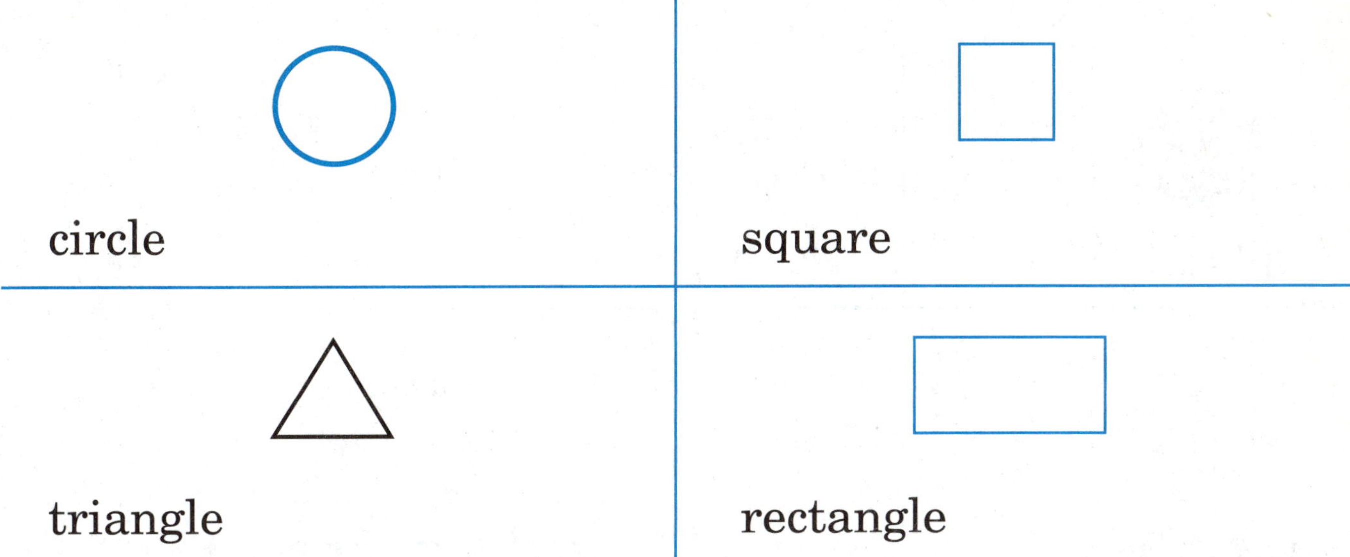

CHAPTER 7

Color all circles blue. Color all squares red. Color all triangles green. Color all rectangles yellow.

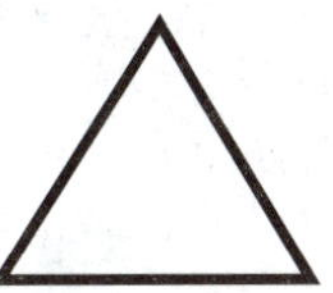

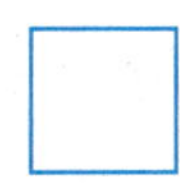

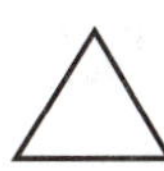

NAME ______________________

Lesson 2 Solid Figures

 sphere	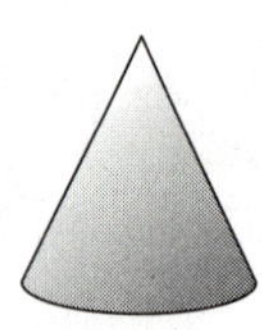 cone	 cube
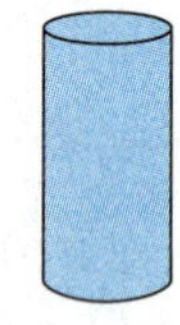 cylinder	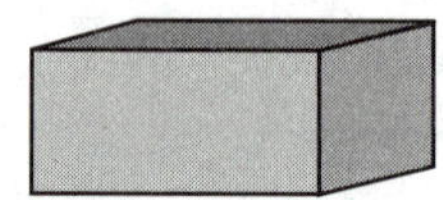 rectangular solid	

Ring the shape.

sphere   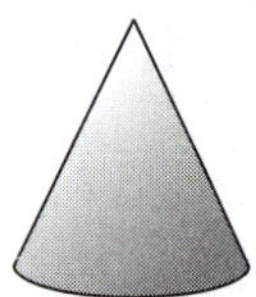

cylinder 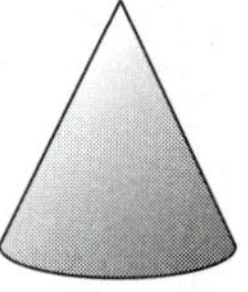 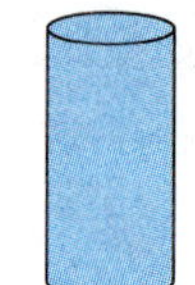 

rectangular solid 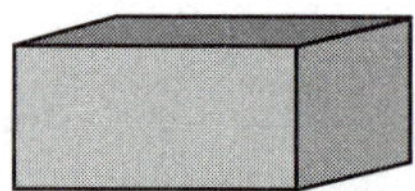  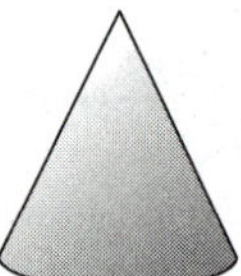

cone   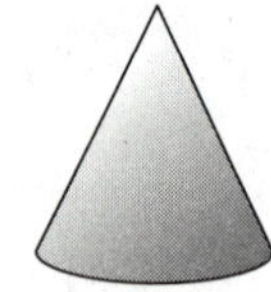

Lesson 2 Solid Figures

Color all cubes blue. Color all cones red. Color all spheres green. Color all rectangular solids yellow. Color all cylinders purple.

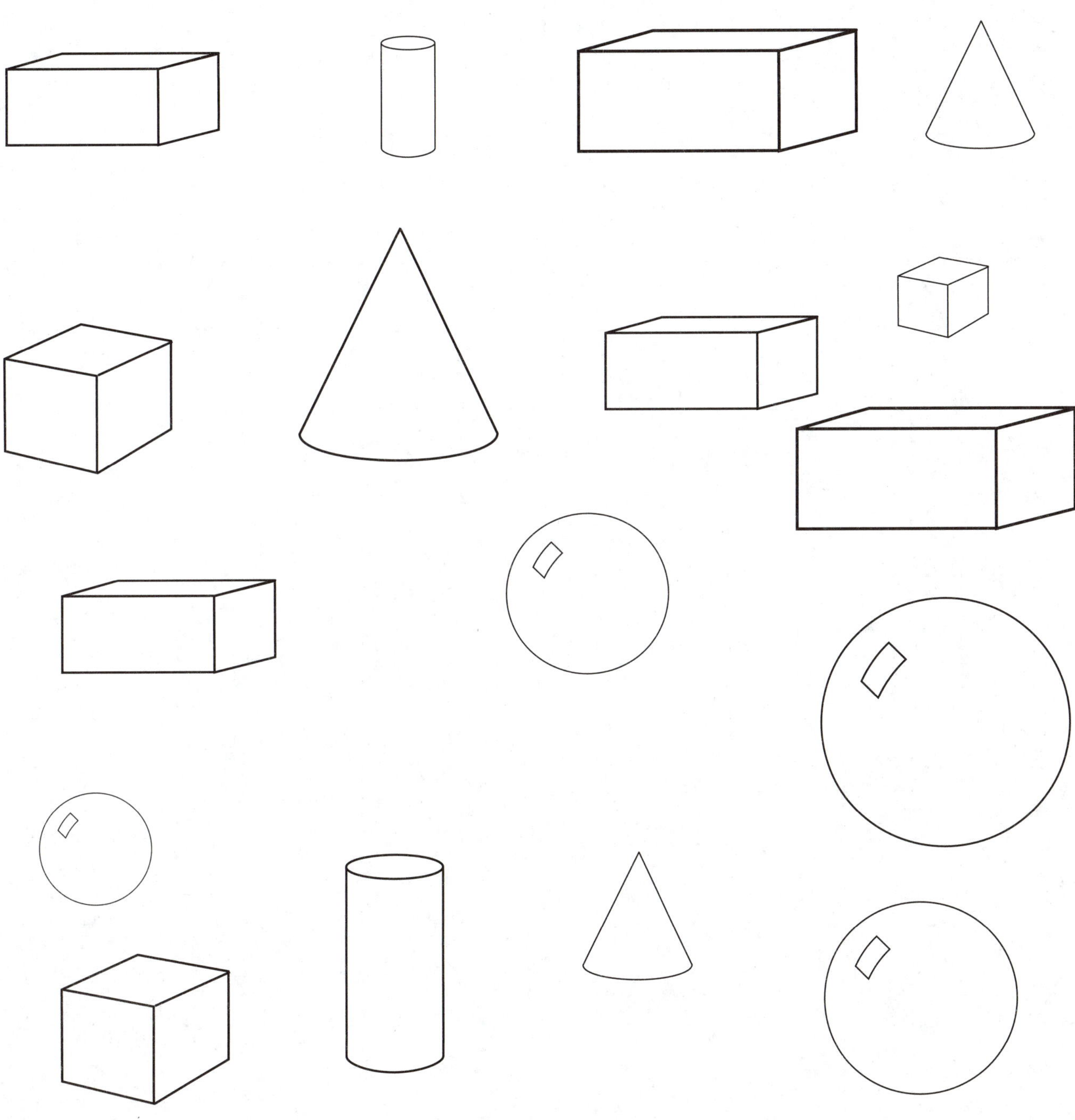

CHAPTER 7

NAME ________________________

Lesson 3 Geometric Patterns

Find the pattern. Ring the shape that comes next.

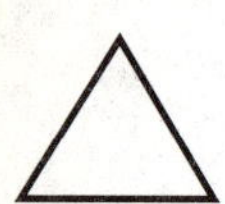  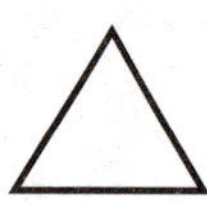    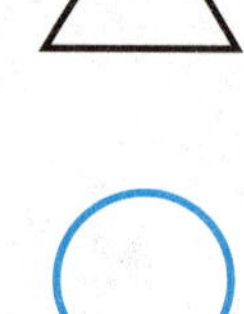

      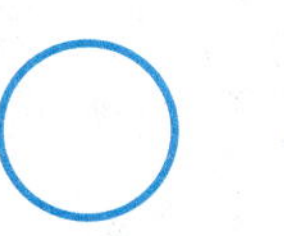  

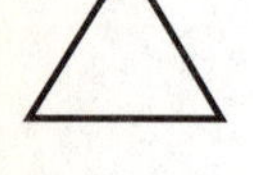   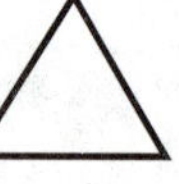   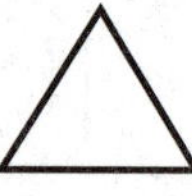

Lesson 4 Geometric Patterns

Find the pattern. Ring the shape that comes next.

  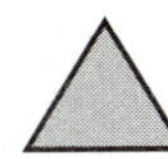    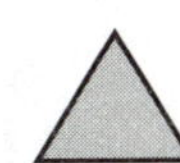 

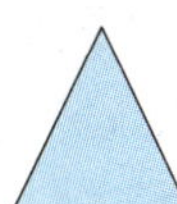 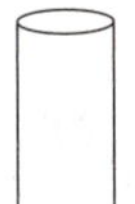 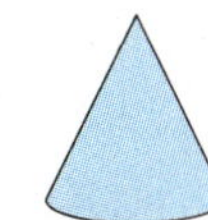 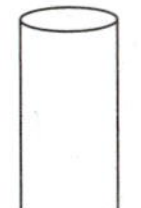 

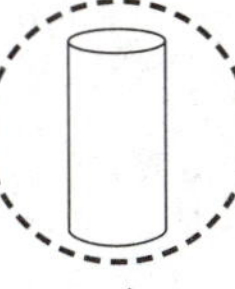
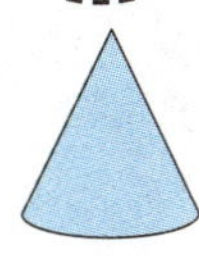

         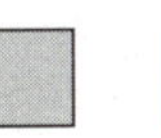 

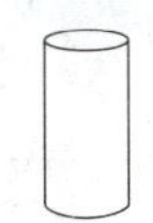

        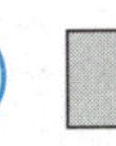 

   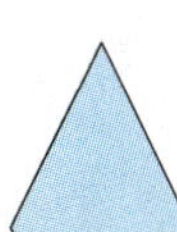    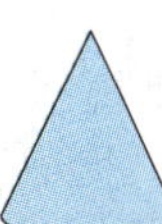

CHAPTER 7

Lesson 5 Symmetry

Ring the pictures that match when folded on the line.

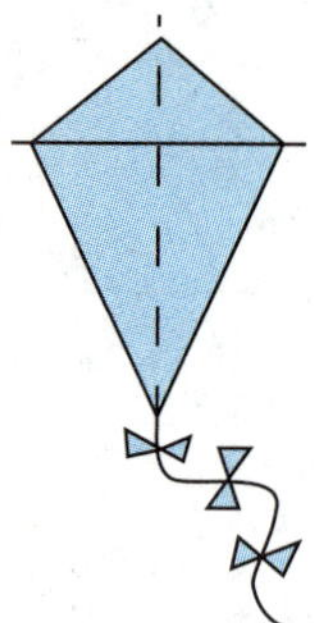
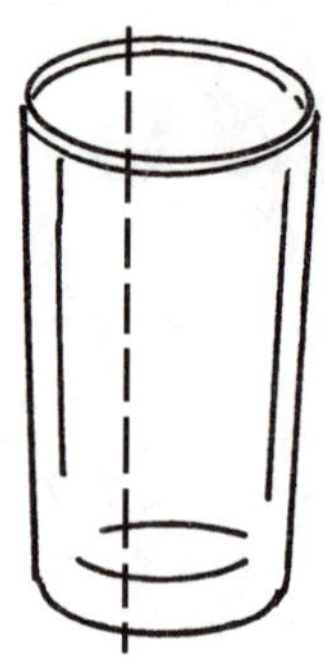

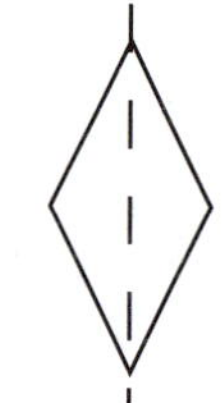
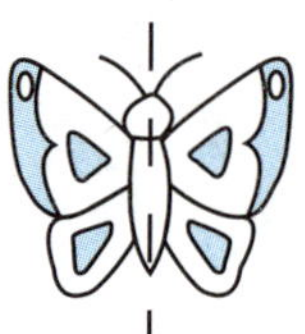

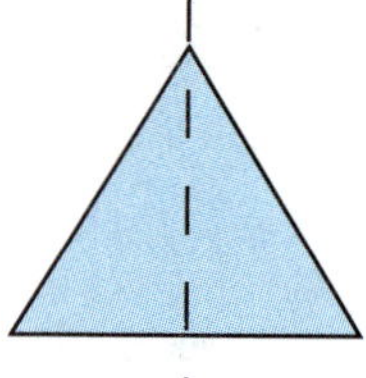

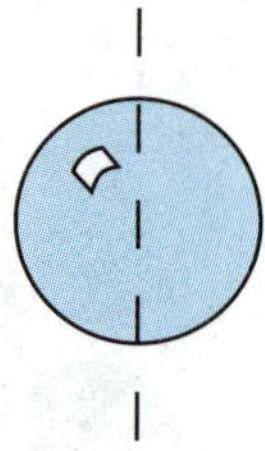

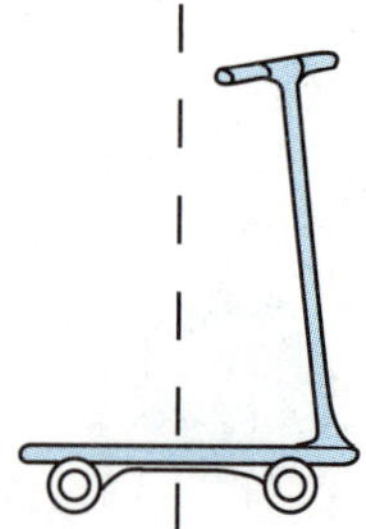

NAME ______________________

CHAPTER 7 PRACTICE TEST
Geometry

Ring the shape.

triangle   

cone 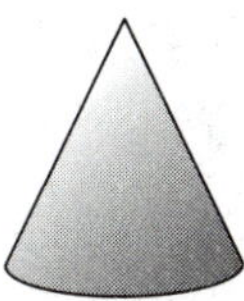 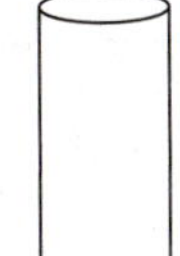 

square 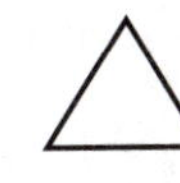  

cube 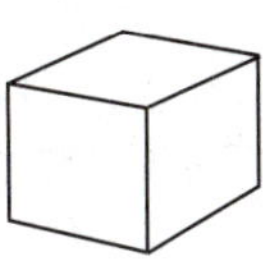  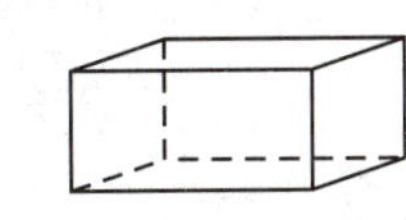

CHAPTER 7

Ring the pictures that match when folded on the line.

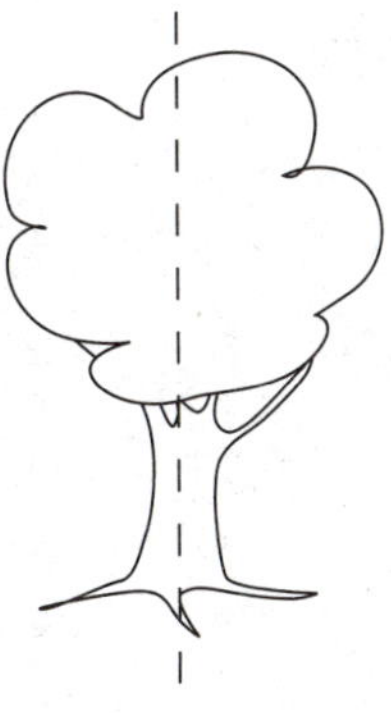

NAME ______________________

CHAPTER 8 PRETEST
Addition and Subtraction (2-digit with no renaming)

Add.

50 +30	26 + 3	44 +12	13 +54	60 +25
24 +62	41 +37	27 +32	38 +10	42 +45
78 +20	42 +33	52 +32	62 +25	46 +52

CHAPTER 8

Subtract.

60 −30	38 − 4	37 −10	26 −12	47 −31
28 − 6	56 −35	68 −33	29 −12	58 −21
67 −20	73 −21	48 −24	39 −15	95 −62

NAME ______________________

Lesson 1 Addition and Subtraction Facts

Add.

5 +2	7 +1	3 +7	2 +2	6 +3	3 +2
1 +8	3 +1	1 +1	2 +8	3 +4	5 +4
2 +1	5 +5	9 +1	2 +6	4 +4	10 +0

Subtract.

8 −1	9 −9	6 −3	7 −0	4 −2	10 −3
10 −9	9 −6	7 −6	10 −8	5 −4	9 −2
5 −3	8 −5	10 −6	6 −4	9 −8	7 −4

Add or subtract.
If you get 5, color
that part green.

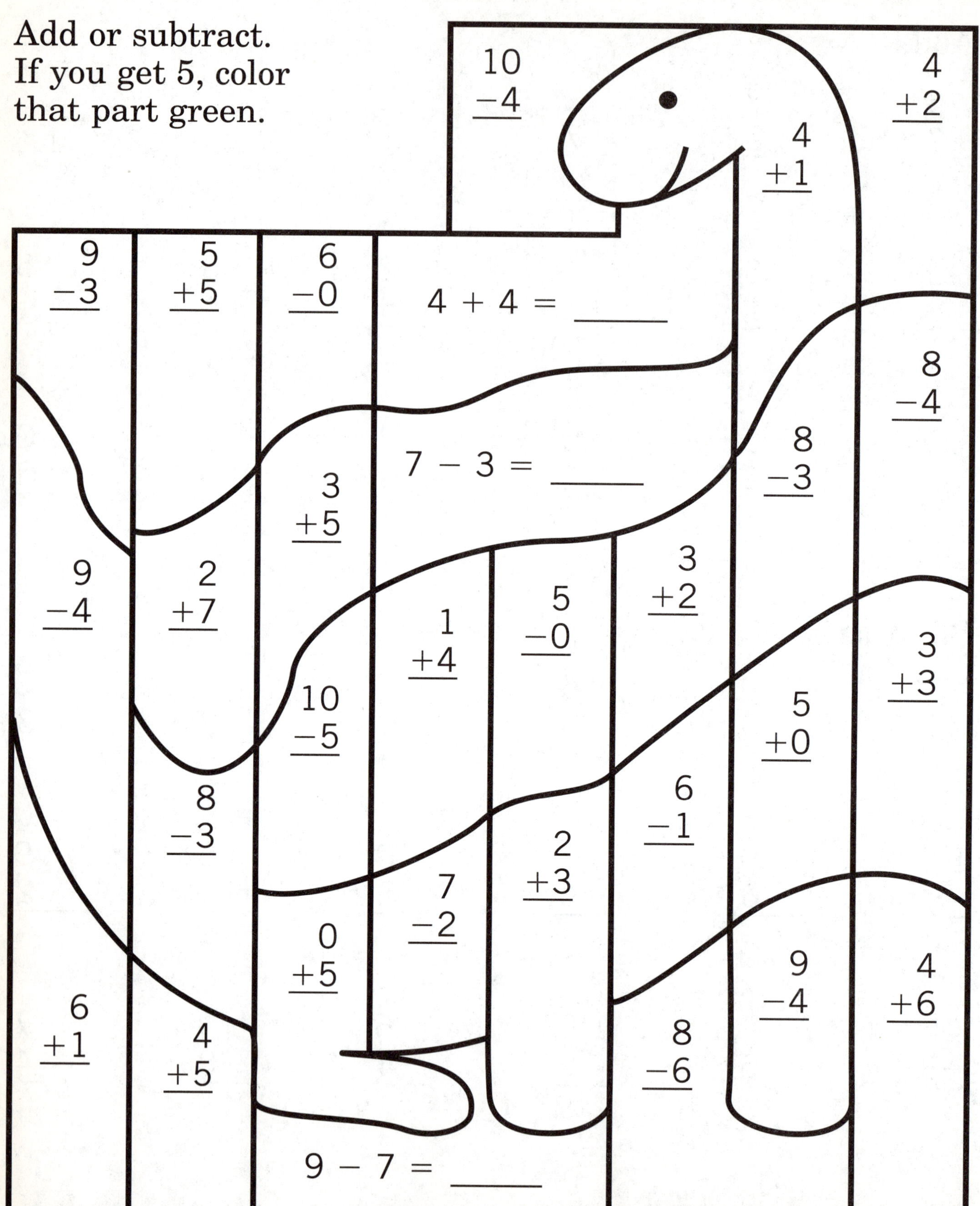

Lesson 2 Adding and Subtracting Tens

$$\begin{array}{r} 2 \text{ tens} \\ +\ 6 \text{ tens} \\ \hline 8 \text{ tens} \end{array} \qquad \begin{array}{r} 20 \\ +60 \\ \hline 80 \end{array} \qquad \begin{array}{r} 8 \text{ tens} \\ -\ 6 \text{ tens} \\ \hline 2 \text{ tens} \end{array} \qquad \begin{array}{r} 80 \\ -60 \\ \hline 20 \end{array}$$

Add.

$$\begin{array}{r} 4 \text{ tens} \\ +\ 5 \text{ tens} \\ \hline \text{tens} \end{array} \qquad \begin{array}{r} 40 \\ +50 \\ \hline \end{array}$$

$$\begin{array}{r} 1 \text{ ten} \\ +\ 5 \text{ tens} \\ \hline \text{tens} \end{array} \qquad \begin{array}{r} 10 \\ +50 \\ \hline \end{array}$$

$$\begin{array}{r} 30 \\ +40 \\ \hline \end{array} \qquad \begin{array}{r} 80 \\ +10 \\ \hline \end{array} \qquad \begin{array}{r} 60 \\ +30 \\ \hline \end{array}$$

$$\begin{array}{r} 20 \\ +30 \\ \hline \end{array} \qquad \begin{array}{r} 40 \\ +40 \\ \hline \end{array} \qquad \begin{array}{r} 10 \\ +70 \\ \hline \end{array}$$

$$\begin{array}{r} 50 \\ +20 \\ \hline \end{array} \qquad \begin{array}{r} 60 \\ +10 \\ \hline \end{array} \qquad \begin{array}{r} 30 \\ +30 \\ \hline \end{array}$$

Subtract.

$$\begin{array}{r} 9 \text{ tens} \\ -\ 5 \text{ tens} \\ \hline \text{tens} \end{array} \qquad \begin{array}{r} 90 \\ -50 \\ \hline \end{array}$$

$$\begin{array}{r} 6 \text{ tens} \\ -\ 5 \text{ tens} \\ \hline \text{ten} \end{array} \qquad \begin{array}{r} 60 \\ -50 \\ \hline \end{array}$$

$$\begin{array}{r} 50 \\ -40 \\ \hline \end{array} \qquad \begin{array}{r} 60 \\ -20 \\ \hline \end{array} \qquad \begin{array}{r} 90 \\ -30 \\ \hline \end{array}$$

$$\begin{array}{r} 20 \\ -10 \\ \hline \end{array} \qquad \begin{array}{r} 90 \\ -70 \\ \hline \end{array} \qquad \begin{array}{r} 30 \\ -20 \\ \hline \end{array}$$

$$\begin{array}{r} 80 \\ -30 \\ \hline \end{array} \qquad \begin{array}{r} 70 \\ -40 \\ \hline \end{array} \qquad \begin{array}{r} 90 \\ -60 \\ \hline \end{array}$$

Lesson 2 Problem Solving

Solve each problem.

There are 90  .

40 are used.

How many are not used?

You have 20  .

You buy 30 more.

Now how many do you have?

Gina had 60  .

She spent 40  .

How many  does she have left?

Zach found 10  .

Then he found 20 more.

Now how many does he have?

There were 80  .

60 ran away.

Then how many were left?

NAME ______________________

Lesson 3 Addition (2-digit)

Join the pennies.
Add the ones.

$$\begin{array}{r} 34 \\ +25 \\ \hline 9 \end{array}$$

Join the dimes.
Add the tens.

$$\begin{array}{r} 34 \\ +25 \\ \hline 59 \end{array}$$

Add.

$$\begin{array}{r} 47 \\ +2 \\ \hline \end{array}$$

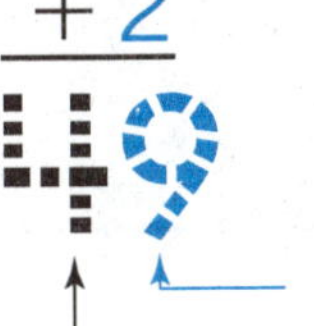

Add the ones.
Add the tens.

$$\begin{array}{r} 52 \\ +44 \\ \hline \end{array} \quad \begin{array}{r} 84 \\ +10 \\ \hline \end{array} \quad \begin{array}{r} 26 \\ +13 \\ \hline \end{array}$$

$$\begin{array}{r} 11 \\ +14 \\ \hline \end{array} \quad \begin{array}{r} 31 \\ +12 \\ \hline \end{array} \quad \begin{array}{r} 12 \\ +1 \\ \hline \end{array} \quad \begin{array}{r} 78 \\ +11 \\ \hline \end{array} \quad \begin{array}{r} 43 \\ +10 \\ \hline \end{array}$$

$$\begin{array}{r} 50 \\ +18 \\ \hline \end{array} \quad \begin{array}{r} 18 \\ +50 \\ \hline \end{array} \quad \begin{array}{r} 81 \\ +5 \\ \hline \end{array} \quad \begin{array}{r} 75 \\ +23 \\ \hline \end{array} \quad \begin{array}{r} 54 \\ +42 \\ \hline \end{array}$$

$$\begin{array}{r} 43 \\ +16 \\ \hline \end{array} \quad \begin{array}{r} 22 \\ +26 \\ \hline \end{array} \quad \begin{array}{r} 43 \\ +2 \\ \hline \end{array} \quad \begin{array}{r} 33 \\ +54 \\ \hline \end{array} \quad \begin{array}{r} 31 \\ +26 \\ \hline \end{array}$$

CHAPTER 8

Lesson 3 Problem Solving

Solve each problem.

There are 24  .

35 more are planted.

Now how many are there?

Al had 27  .

He bought 12 more.

Now how many does he have?

Ima has 54  .

Max has 34  .

How many do they have in all?

You found 82  .

Then you find 7 more.

Now how many do you have?

---

20  are blue.

79  are white.

How many  and  are there in all?

NAME ____________________

Lesson 4 Addition (2-digit)

Add.

24 +13 37	75 + 4 79	50 +27	62 +15	46 +23
52 +34	96 + 2	73 +16	38 +40	35 +21
10 +21	14 + 5	12 +34	33 +53	13 +11
24 +21	57 + 2	60 +33	12 +43	71 +26
16 +52	28 + 1	51 +27	40 +45	63 +16
22 +67	64 + 4	24 +72	41 +38	31 +56

CHAPTER 8

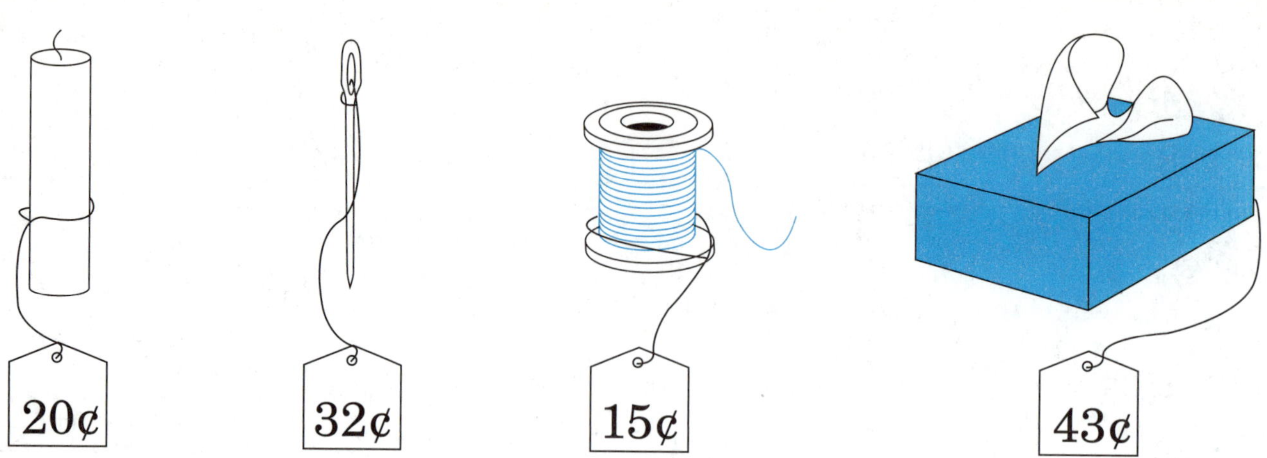

Solve each problem.

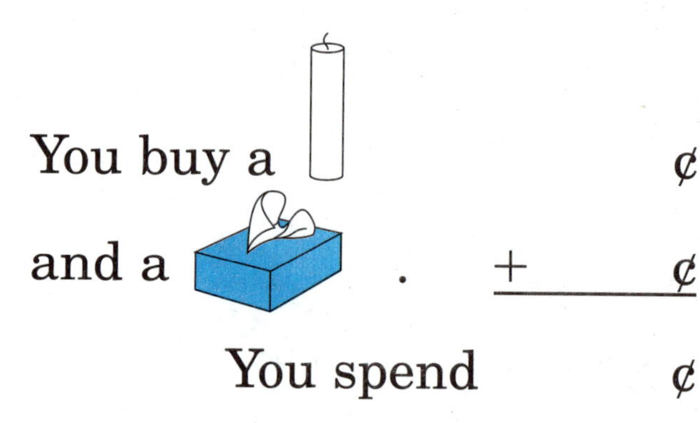

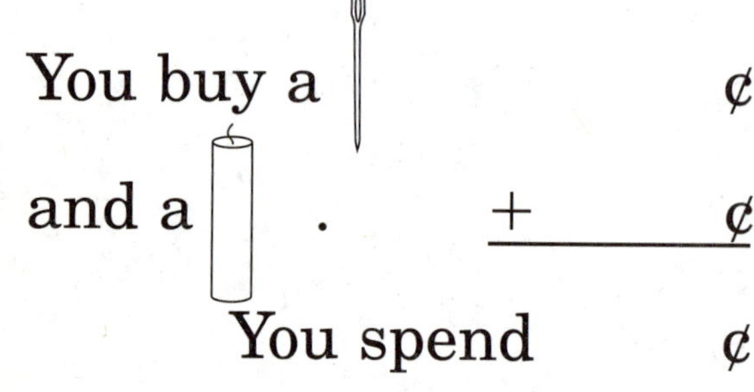

NAME ____________

Lesson 5 Subtraction (2-digit)

Take away 4 pennies.
Subtract the ones.

$$\begin{array}{r} 36 \\ -24 \\ \hline 2 \end{array}$$

Take away 2 dimes.
Subtract the tens.

$$\begin{array}{r} 36 \\ -24 \\ \hline 12 \end{array}$$

CHAPTER 8

Subtract.

$$\begin{array}{r} 78 \\ -6 \\ \hline \end{array}$$

Subtract the ones.
Subtract the tens.

		$\begin{array}{r} 69 \\ -47 \\ \hline \end{array}$	$\begin{array}{r} 28 \\ -15 \\ \hline \end{array}$	$\begin{array}{r} 45 \\ -32 \\ \hline \end{array}$
$\begin{array}{r} 59 \\ -45 \\ \hline \end{array}$	$\begin{array}{r} 98 \\ -43 \\ \hline \end{array}$	$\begin{array}{r} 17 \\ -5 \\ \hline \end{array}$	$\begin{array}{r} 57 \\ -43 \\ \hline \end{array}$	$\begin{array}{r} 48 \\ -34 \\ \hline \end{array}$
$\begin{array}{r} 58 \\ -17 \\ \hline \end{array}$	$\begin{array}{r} 85 \\ -25 \\ \hline \end{array}$	$\begin{array}{r} 87 \\ -7 \\ \hline \end{array}$	$\begin{array}{r} 96 \\ -80 \\ \hline \end{array}$	$\begin{array}{r} 66 \\ -51 \\ \hline \end{array}$
$\begin{array}{r} 94 \\ -41 \\ \hline \end{array}$	$\begin{array}{r} 39 \\ -22 \\ \hline \end{array}$	$\begin{array}{r} 33 \\ -2 \\ \hline \end{array}$	$\begin{array}{r} 65 \\ -22 \\ \hline \end{array}$	$\begin{array}{r} 78 \\ -65 \\ \hline \end{array}$

Solve each problem.

There are 54  .

32 were picked.

How many are left?

You have 48 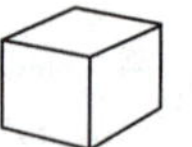 .

35 are blue.

The rest are white.

How many white ones are there?

A store has 99 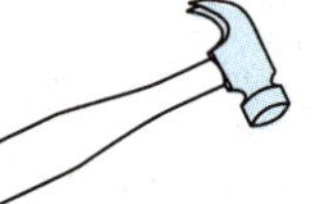 .

The store sold 73.

How many 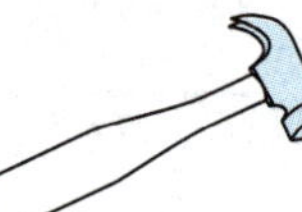 does the store have now?

Yoko has 23 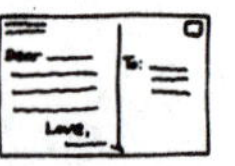 .

Neal has 12  .

How many more  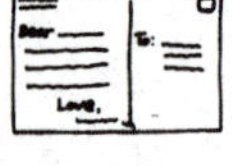 does Yoko have?

To build a you need 48 .

You have 28  .

How many more  do you need?

NAME ______________________

Lesson 6 Subtraction (2-digit)

Subtract.

75 − 34 = 41	67 − 4 = 63	30 − 20	48 − 30	55 − 32
78 − 67	56 − 3	98 − 86	86 − 15	98 − 48
95 − 31	84 − 2	65 − 45	79 − 48	84 − 50
42 − 10	39 − 6	89 − 42	67 − 21	66 − 36
98 − 73	72 − 2	43 − 13	57 − 32	69 − 15
32 − 11	97 − 5	78 − 22	99 − 16	87 − 47

CHAPTER 8

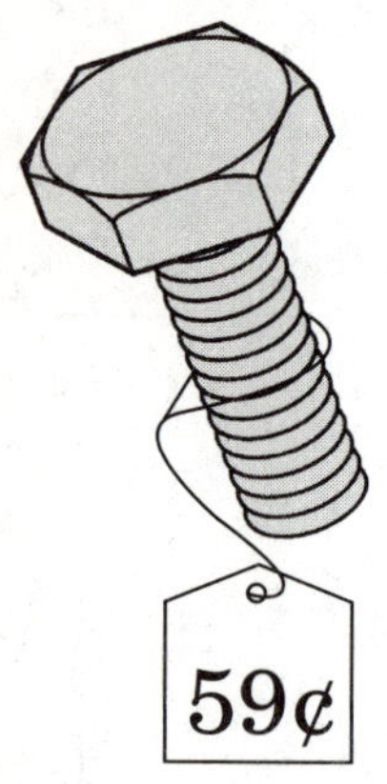

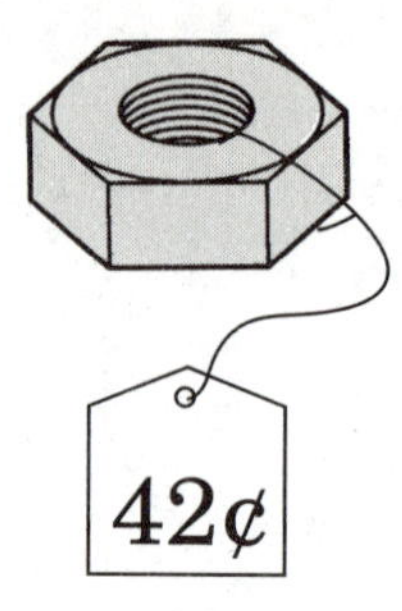

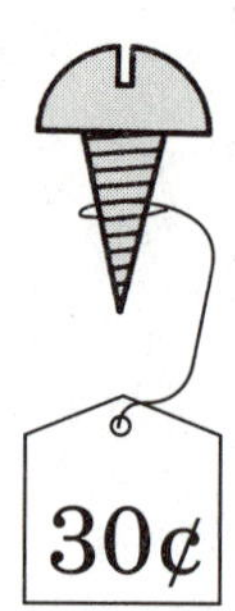

Solve each problem.

You have 6 2¢
You buy a  − 42¢
You have left 20¢

You have 5 0¢
You buy a − 30¢
You have left ¢

You have 5 3¢
You buy a − ¢
You have left ¢

You have 9 9¢
You buy a − ¢
You have left ¢

You have 7 6¢
You buy a − ¢
You have left ¢

You have 4 5¢
You buy a − ¢
You have left ¢

NAME ____________________

Lesson 7 Adding 3 Numbers

Add the ones.

$$\begin{array}{r} 12 \\ 53 \\ +24 \\ \hline 9 \end{array} \quad \begin{array}{r} 5 \\ +4 \\ \hline \end{array}$$

Add the tens.

$$\begin{array}{r} 60 \\ +20 \\ \hline \end{array} \quad \begin{array}{r} 12 \\ 53 \\ +24 \\ \hline 89 \end{array}$$

Add.

$$\begin{array}{r} 45 \\ 33 \\ +10 \\ \hline 88 \end{array}$$

Add the ones.
Add the tens.

$$\begin{array}{r} 32 \\ 37 \\ +20 \\ \hline 89 \end{array} \qquad \begin{array}{r} 35 \\ 21 \\ +11 \\ \hline \end{array} \qquad \begin{array}{r} 47 \\ 20 \\ +22 \\ \hline \end{array}$$

CHAPTER 8

$$\begin{array}{r} 24 \\ 24 \\ +21 \\ \hline \end{array} \qquad \begin{array}{r} 31 \\ 23 \\ +31 \\ \hline \end{array} \qquad \begin{array}{r} 40 \\ 13 \\ +11 \\ \hline \end{array} \qquad \begin{array}{r} 54 \\ 10 \\ +23 \\ \hline \end{array} \qquad \begin{array}{r} 25 \\ 33 \\ +20 \\ \hline \end{array}$$

$$\begin{array}{r} 61 \\ 12 \\ +24 \\ \hline \end{array} \qquad \begin{array}{r} 30 \\ 24 \\ +15 \\ \hline \end{array} \qquad \begin{array}{r} 36 \\ 32 \\ +31 \\ \hline \end{array} \qquad \begin{array}{r} 31 \\ 20 \\ +24 \\ \hline \end{array} \qquad \begin{array}{r} 44 \\ 20 \\ +34 \\ \hline \end{array}$$

Lesson 7 Problem Solving

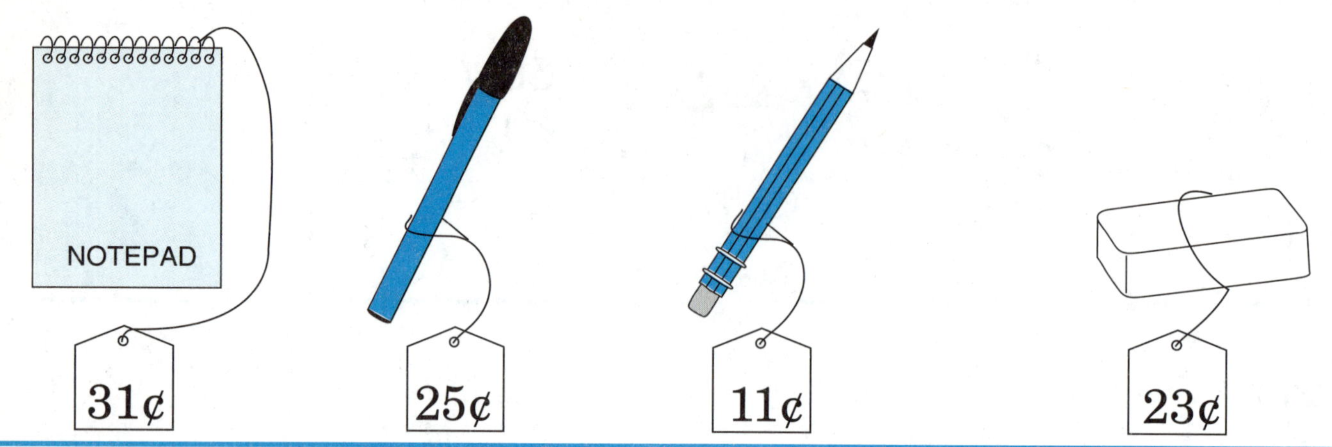

Solve each problem.

You buy a 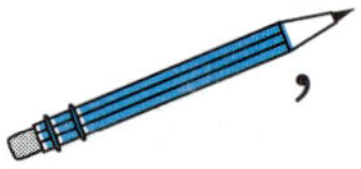 , 11 ¢
an  , 23 ¢
and a  . +31 ¢
You spend ¢

You buy a NOTEPAD , ¢
a  , ¢
and a  . + ¢
You spend ¢

You buy an 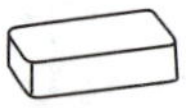 , ¢
a NOTEPAD , ¢
and a  . + ¢
You spend ¢

You buy a  , ¢
a  , ¢
and an 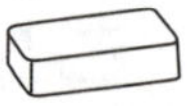 . + ¢
You spend ¢

Lesson 8 Addition and Subtraction

Add or subtract. 

Watch the + and − !

$\begin{array}{r} 32 \\ +34 \\ \hline \end{array}$	$\begin{array}{r} 57 \\ -25 \\ \hline \end{array}$	$\begin{array}{r} 82 \\ -\ \ 2 \\ \hline \end{array}$	$\begin{array}{r} 92 \\ -61 \\ \hline \end{array}$	$\begin{array}{r} 42 \\ +57 \\ \hline \end{array}$
$\begin{array}{r} 86 \\ -52 \\ \hline \end{array}$	$\begin{array}{r} 15 \\ +62 \\ \hline \end{array}$	$\begin{array}{r} 40 \\ +\ \ 7 \\ \hline \end{array}$	$\begin{array}{r} 31 \\ +15 \\ \hline \end{array}$	$\begin{array}{r} 73 \\ -13 \\ \hline \end{array}$
$\begin{array}{r} 69 \\ +30 \\ \hline \end{array}$	$\begin{array}{r} 34 \\ -12 \\ \hline \end{array}$	$\begin{array}{r} 87 \\ -\ \ 6 \\ \hline \end{array}$	$\begin{array}{r} 84 \\ +14 \\ \hline \end{array}$	$\begin{array}{r} 40 \\ +35 \\ \hline \end{array}$
$\begin{array}{r} 95 \\ -25 \\ \hline \end{array}$	$\begin{array}{r} 86 \\ -43 \\ \hline \end{array}$	$\begin{array}{r} 77 \\ +\ \ 2 \\ \hline \end{array}$	$\begin{array}{r} 61 \\ +17 \\ \hline \end{array}$	$\begin{array}{r} 59 \\ -44 \\ \hline \end{array}$
$\begin{array}{r} 39 \\ +60 \\ \hline \end{array}$	$\begin{array}{r} 68 \\ -67 \\ \hline \end{array}$	$\begin{array}{r} 55 \\ +\ \ 4 \\ \hline \end{array}$	$\begin{array}{r} 88 \\ -83 \\ \hline \end{array}$	$\begin{array}{r} 96 \\ -81 \\ \hline \end{array}$

Lesson 8 Problem Solving

Solve each problem.

There are 43 white .

There are 45 blue .

How many more blue  than white  are there?

$$\begin{array}{r} 45 \\ -43 \\ \hline \end{array}$$

There are 19 .

30 more come.

How many are there now?

---

34  are on the table.

2 are broken.

How many are not broken?

You need 19 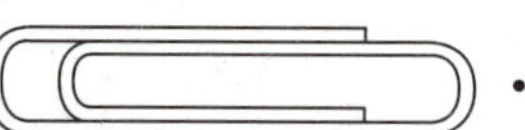.

You have 7 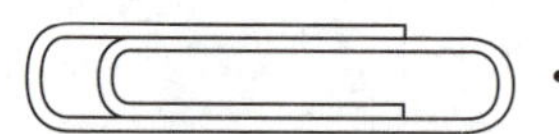.

How many more do you need?

You have 32 .

You get 57 more.

Now how many do you have?

NAME ____________________

Lesson 9 Addition and Subtraction

Add or subtract. STOP! Watch the + and − !

If you get 56, color the box.

22 +34 = 56	78 −24	59 − 3	75 −11	15 +41
96 −40	15 +62	50 + 6	61 +15	67 −11
69 −13	78 −22	57 − 1	84 +14	40 +16
97 −41	86 −32	54 + 2	61 +17	99 −43
31 +25	68 −67	52 + 4	88 −83	79 −23

CHAPTER 8

Addition and Subtraction

Add or subtract.  Watch the + and − !

If you get 75, color the box.

71 + 4 75	83 +14	44 +34	63 +12	73 + 2
70 + 5	32 +43	97 −22	44 +31	76 − 1
87 −12	90 + 7	85 −22	61 +25	79 − 4
95 −20	78 − 3	85 −10	61 +14	50 +25
72 + 3	85 +10	51 −20	67 −30	95 −20

NAME ________________________

CHAPTER 8 PRACTICE TEST

Addition and Subtraction (2-digit with no renaming)

Add.

$$\begin{array}{r} 30 \\ +50 \\ \hline \end{array} \qquad \begin{array}{r} 34 \\ +\ \ 5 \\ \hline \end{array} \qquad \begin{array}{r} 61 \\ +24 \\ \hline \end{array} \qquad \begin{array}{r} 18 \\ +31 \\ \hline \end{array} \qquad \begin{array}{r} 50 \\ +36 \\ \hline \end{array}$$

$$\begin{array}{r} 45 \\ +21 \\ \hline \end{array} \qquad \begin{array}{r} 92 \\ +\ \ 7 \\ \hline \end{array} \qquad \begin{array}{r} 73 \\ +13 \\ \hline \end{array} \qquad \begin{array}{r} 54 \\ +24 \\ \hline \end{array} \qquad \begin{array}{r} 82 \\ +17 \\ \hline \end{array}$$

$$\begin{array}{r} 35 \\ +12 \\ \hline \end{array} \qquad \begin{array}{r} 60 \\ +\ \ 8 \\ \hline \end{array} \qquad \begin{array}{r} 30 \\ 40 \\ +10 \\ \hline \end{array} \qquad \begin{array}{r} 51 \\ 16 \\ +12 \\ \hline \end{array} \qquad \begin{array}{r} 26 \\ 21 \\ +12 \\ \hline \end{array}$$

CHAPTER 8

Subtract.

$$\begin{array}{r} 70 \\ -40 \\ \hline \end{array} \qquad \begin{array}{r} 85 \\ -\ \ 2 \\ \hline \end{array} \qquad \begin{array}{r} 54 \\ -30 \\ \hline \end{array} \qquad \begin{array}{r} 97 \\ -35 \\ \hline \end{array} \qquad \begin{array}{r} 74 \\ -54 \\ \hline \end{array}$$

$$\begin{array}{r} 28 \\ -18 \\ \hline \end{array} \qquad \begin{array}{r} 78 \\ -\ \ 4 \\ \hline \end{array} \qquad \begin{array}{r} 46 \\ -23 \\ \hline \end{array} \qquad \begin{array}{r} 89 \\ -33 \\ \hline \end{array} \qquad \begin{array}{r} 93 \\ -62 \\ \hline \end{array}$$

CHAPTER 9 PRETEST
Addition and Subtraction (facts through 18)

Add.

8 + 9	7 + 5	7 + 7	6 + 5	9 + 6	7 + 9
6 + 7	8 + 4	9 + 2	4 + 9	8 + 8	9 + 5
7 + 4	5 + 9	6 + 8	8 + 5	9 + 4	9 + 8

Subtract.

11 − 4	17 − 8	14 − 8	13 − 5	13 − 4	19 − 14
13 − 6	12 − 4	11 − 2	13 − 9	16 − 8	14 − 5
17 − 9	16 − 9	14 − 7	15 − 6	12 − 5	11 − 5

NAME ____________

Lesson 1 Facts for 11

Add or subtract.

8 +3 11		3 +8
11 − 8 3		11 − 3
9 +2		2 +9
11 − 9		11 − 2
6 +5		5 +6
11 − 6		11 − 5
7 +4		4 +7
11 − 7		11 − 4

7 +4	6 +5	3 +8	11 − 6	11 − 9	11 − 7
11 +0	9 +2	5 +6	11 − 2	11 − 0	11 − 8

CHAPTER 9

Lesson 1 Problem Solving

Solve each problem.

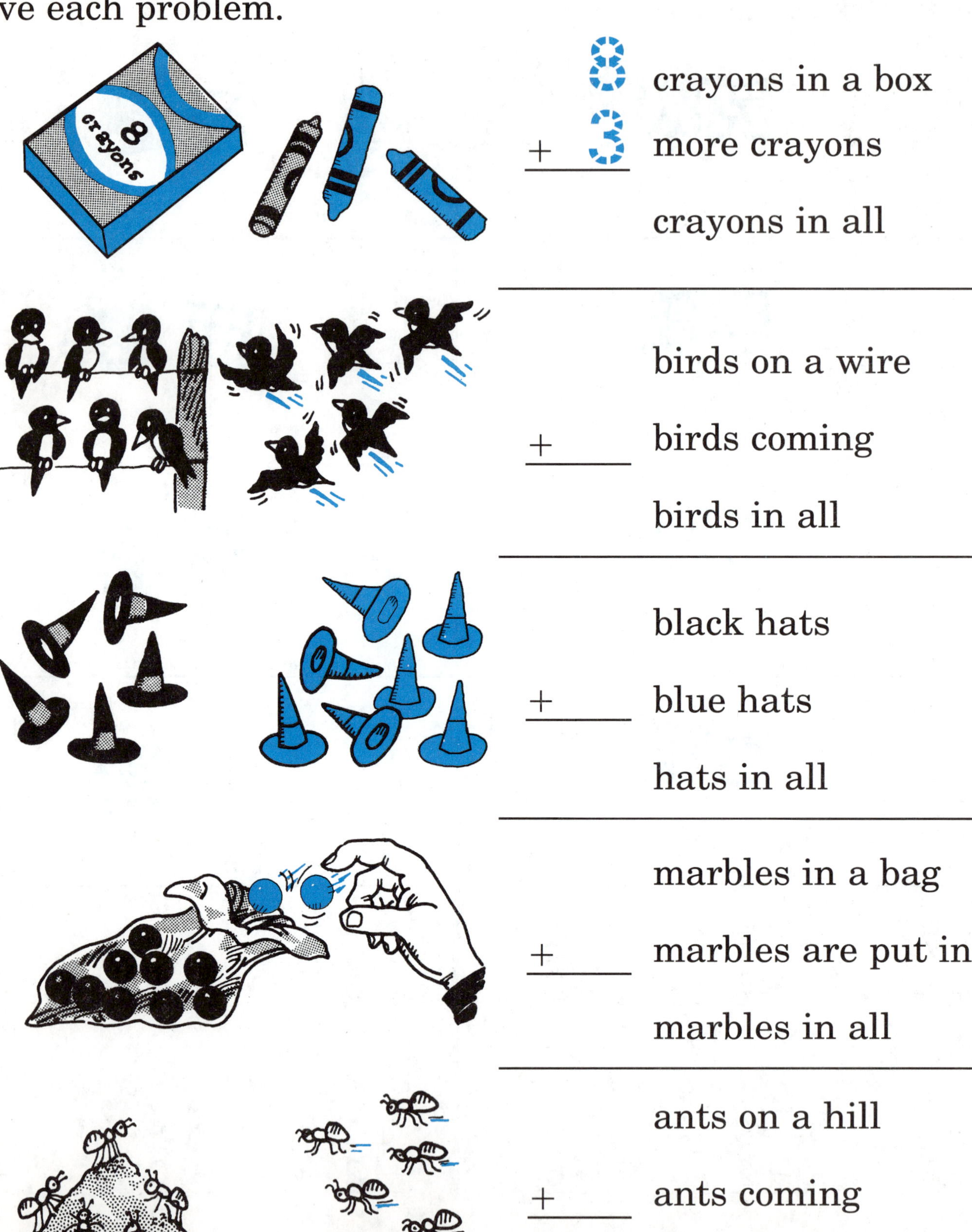

8 crayons in a box

+ 3 more crayons

___ crayons in all

___ birds on a wire

+ ___ birds coming

___ birds in all

___ black hats

+ ___ blue hats

___ hats in all

___ marbles in a bag

+ ___ marbles are put in

___ marbles in all

___ ants on a hill

+ ___ ants coming

___ ants in all

Lesson 2 Facts for 12

Add or subtract.

$\begin{array}{r} 8 \\ +4 \\ \hline 12 \end{array}$  $\begin{array}{r} 4 \\ +8 \\ \hline \end{array}$ $\begin{array}{r} 12 \\ -8 \\ \hline 4 \end{array}$ $\begin{array}{r} 12 \\ -4 \\ \hline \end{array}$

$\begin{array}{r} 9 \\ +3 \\ \hline \end{array}$  $\begin{array}{r} 3 \\ +9 \\ \hline \end{array}$ $\begin{array}{r} 12 \\ -9 \\ \hline \end{array}$ $\begin{array}{r} 12 \\ -3 \\ \hline \end{array}$

$\begin{array}{r} 7 \\ +5 \\ \hline \end{array}$  $\begin{array}{r} 5 \\ +7 \\ \hline \end{array}$ $\begin{array}{r} 12 \\ -7 \\ \hline \end{array}$ $\begin{array}{r} 12 \\ -5 \\ \hline \end{array}$

$\begin{array}{r} 6 \\ +6 \\ \hline \end{array}$  $\begin{array}{r} 12 \\ -6 \\ \hline \end{array}$ 

$\begin{array}{r} 9 \\ +3 \\ \hline \end{array}$ $\begin{array}{r} 5 \\ +7 \\ \hline \end{array}$ $\begin{array}{r} 12 \\ +0 \\ \hline \end{array}$ $\begin{array}{r} 12 \\ -6 \\ \hline \end{array}$ $\begin{array}{r} 12 \\ -0 \\ \hline \end{array}$ $\begin{array}{r} 12 \\ -8 \\ \hline \end{array}$

$\begin{array}{r} 6 \\ +6 \\ \hline \end{array}$ $\begin{array}{r} 4 \\ +8 \\ \hline \end{array}$ $\begin{array}{r} 7 \\ +5 \\ \hline \end{array}$ $\begin{array}{r} 12 \\ -5 \\ \hline \end{array}$ $\begin{array}{r} 12 \\ -9 \\ \hline \end{array}$ $\begin{array}{r} 12 \\ -4 \\ \hline \end{array}$

Lesson 2 Problem Solving

Solve each problem.

12 birds in all

− 4 birds flying away

birds stay

cars in all

− cars leaving

cars stay

flowers in all

− blue flowers

white flowers

BUTTONS

blue buttons

+ black buttons

buttons in all

berries in all

− berries falling

berries not falling

NAME ____________

Lesson 3 Facts through 12

Add.

8 +4 12	7 +4	6 +6	3 +9	9 +2	6 +5
5 +5	7 +5	9 +1	4 +8	5 +7	3 +8
4 +6	2 +9	8 +3	4 +7	9 +3	5 +6

CHAPTER 9

Subtract.

12 − 8 4	11 − 9	12 − 5	11 − 4	12 − 6	11 − 0
12 − 3	12 − 7	10 − 3	11 − 8	10 − 6	11 − 5
12 − 0	11 − 2	10 − 8	12 − 4	11 − 7	12 − 9

Lesson 3 Problem Solving

Solve each problem.

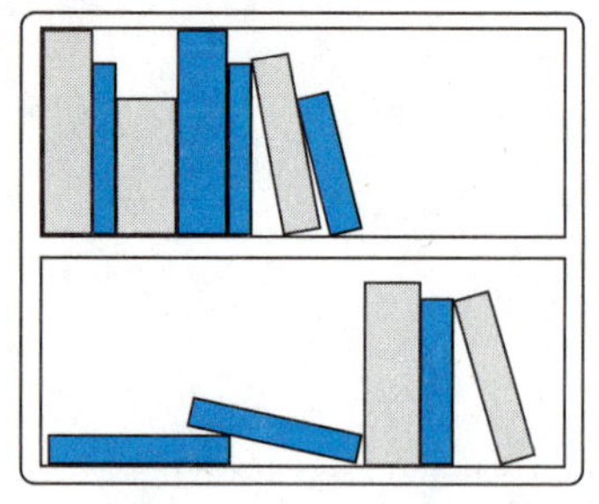

	books on top shelf
+ ____	books on bottom shelf
	books in all

	cars in all
− ____	cars going
	cars still parked

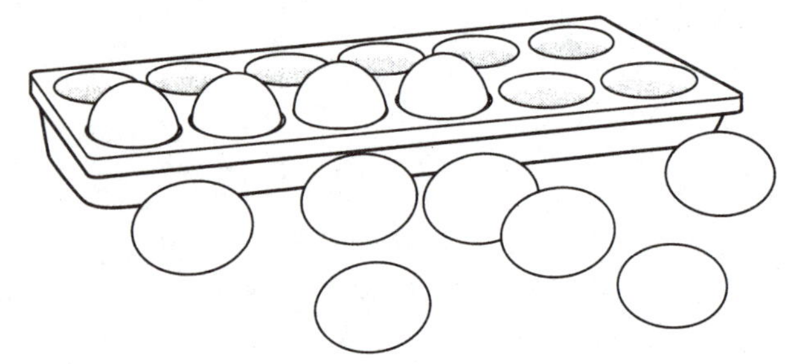

	eggs in the carton
+ ____	eggs on the table
	eggs in all

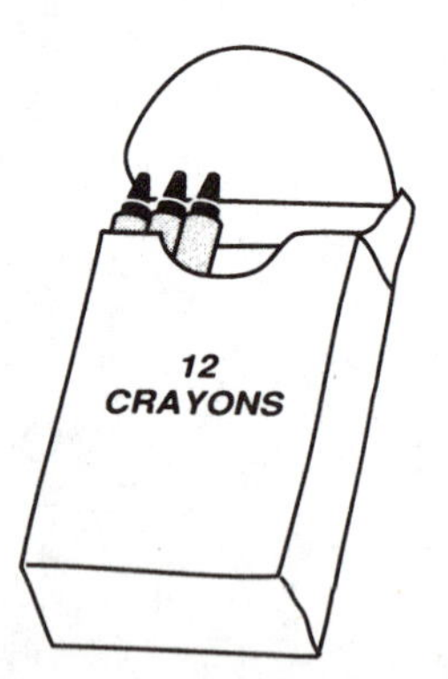

	crayons that can be in the box
− ____	crayons in the box
	crayons needed to fill the box

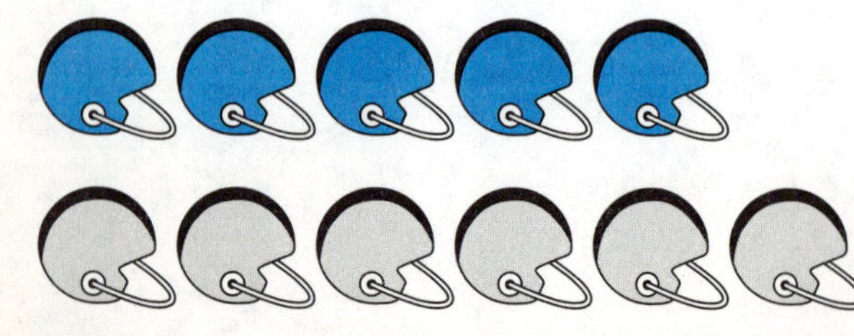

	blue football helmets
+ ____	gray football helmets
	helmets in all

Lesson 4 Facts for 13

Add or subtract.

7 +6 = 13		6 +7	13 −7 = 6		13 − 6
8 +5		5 +8	13 − 8		13 − 5
9 +4		4 +9	13 − 9		13 − 4
5 +8	9 +4	7 +6	13 − 5	13 − 7	13 − 9
6 +7	8 +5	4 +9	13 − 4	13 − 8	13 − 6

CHAPTER 9

Lesson 5 Facts for 14

Add or subtract.

9 +5 14		5 +9	
14 − 9 5		14 − 5	
8 +6		6 +8	
14 − 8		14 − 6	
7 +7			
14 −7			

8 +6	7 +7	6 +6	14 − 8	12 − 7	14 − 5
9 +4	5 +9	7 +4	14 − 7	13 − 8	14 − 6
9 +5	6 +7	6 +8	11 − 8	14 − 9	12 − 4

Lesson 6 Facts through 14

Add.

9	7	6	8	5	4
+4	+7	+5	+4	+9	+7
13					

5	9	5	8	7	6
+6	+5	+7	+5	+6	+8

3	7	8	5	4	6
+8	+5	+6	+8	+9	+6

Subtract.

12	13	14	11	14	13
− 5	− 7	− 5	− 3	− 7	− 6
7					

13	12	14	11	14	12
− 8	− 9	− 6	− 5	− 9	− 6

11	11	13	12	13	14
− 7	− 9	− 4	− 8	− 9	− 8

Lesson 6 Problem Solving

Solve each problem.

8 trucks in a box
+ 6 other trucks
___ trucks in all

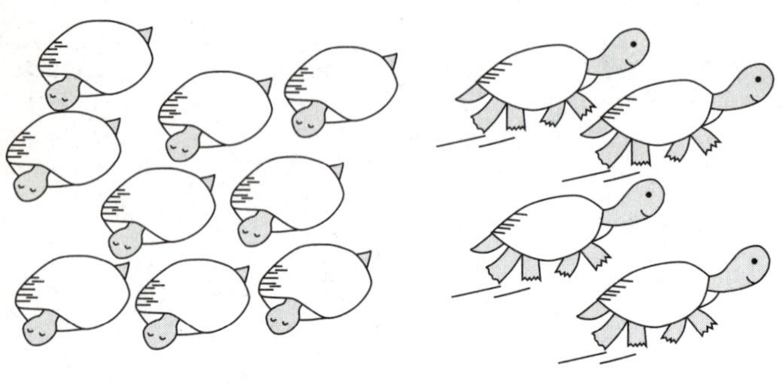

___ turtles in all
− ___ turtles going away
___ turtles stay

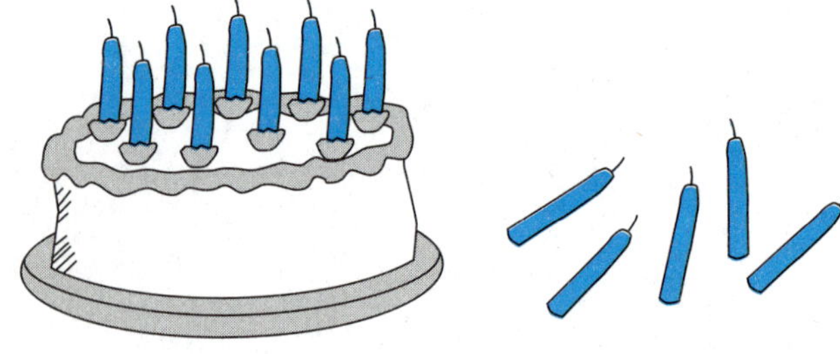

___ candles on the cake
+ ___ other candles
___ candles in all

___ marbles in a bag
+ ___ more marbles
___ marbles in all

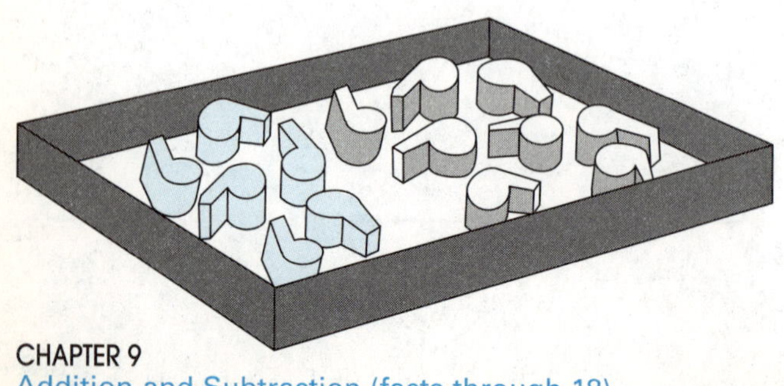

___ whistles in the box
− ___ white whistles
___ blue whistles

NAME ______________________

Lesson 7 Facts for 15

Add or subtract.

* * * * * * * *
* * * * * * *

$\begin{array}{r} 8 \\ +7 \\ \hline 15 \end{array}$ $\begin{array}{r} 7 \\ +8 \\ \hline \end{array}$ $\begin{array}{r} 15 \\ -7 \\ \hline 8 \end{array}$ $\begin{array}{r} 15 \\ -8 \\ \hline \end{array}$

* * * * * * * *
* * * * * * *

$\begin{array}{r} 9 \\ +6 \\ \hline \end{array}$ $\begin{array}{r} 6 \\ +9 \\ \hline \end{array}$ $\begin{array}{r} 15 \\ -6 \\ \hline \end{array}$ $\begin{array}{r} 15 \\ -9 \\ \hline \end{array}$

CHAPTER 9

$\begin{array}{r} 6 \\ +9 \\ \hline \end{array}$ $\begin{array}{r} 7 \\ +8 \\ \hline \end{array}$ $\begin{array}{r} 8 \\ +6 \\ \hline \end{array}$ $\begin{array}{r} 14 \\ -9 \\ \hline \end{array}$ $\begin{array}{r} 15 \\ -8 \\ \hline \end{array}$ $\begin{array}{r} 13 \\ -7 \\ \hline \end{array}$

$\begin{array}{r} 6 \\ +8 \\ \hline \end{array}$ $\begin{array}{r} 9 \\ +5 \\ \hline \end{array}$ $\begin{array}{r} 8 \\ +7 \\ \hline \end{array}$ $\begin{array}{r} 15 \\ -9 \\ \hline \end{array}$ $\begin{array}{r} 14 \\ -6 \\ \hline \end{array}$ $\begin{array}{r} 15 \\ -7 \\ \hline \end{array}$

$\begin{array}{r} 8 \\ +5 \\ \hline \end{array}$ $\begin{array}{r} 9 \\ +6 \\ \hline \end{array}$ $\begin{array}{r} 7 \\ +7 \\ \hline \end{array}$ $\begin{array}{r} 13 \\ -8 \\ \hline \end{array}$ $\begin{array}{r} 15 \\ -6 \\ \hline \end{array}$ $\begin{array}{r} 14 \\ -5 \\ \hline \end{array}$

NAME ________________

Lesson 8 Facts for 16

Add or subtract.

$$\begin{array}{r} 9 \\ +7 \\ \hline \end{array} \quad \begin{array}{r} 7 \\ +9 \\ \hline \end{array} \quad \begin{array}{r} 16 \\ -7 \\ \hline \end{array} \quad \begin{array}{r} 16 \\ -9 \\ \hline \end{array}$$

$$\begin{array}{r} 8 \\ +8 \\ \hline \end{array} \quad \begin{array}{r} 16 \\ -8 \\ \hline \end{array}$$

$$\begin{array}{r} 8 \\ +7 \\ \hline \end{array} \quad \begin{array}{r} 9 \\ +7 \\ \hline \end{array} \quad \begin{array}{r} 7 \\ +8 \\ \hline \end{array} \quad \begin{array}{r} 16 \\ -9 \\ \hline \end{array} \quad \begin{array}{r} 15 \\ -7 \\ \hline \end{array} \quad \begin{array}{r} 14 \\ -5 \\ \hline \end{array}$$

$$\begin{array}{r} 8 \\ +8 \\ \hline \end{array} \quad \begin{array}{r} 5 \\ +9 \\ \hline \end{array} \quad \begin{array}{r} 7 \\ +7 \\ \hline \end{array} \quad \begin{array}{r} 14 \\ -9 \\ \hline \end{array} \quad \begin{array}{r} 15 \\ -6 \\ \hline \end{array} \quad \begin{array}{r} 16 \\ -8 \\ \hline \end{array}$$

$$\begin{array}{r} 7 \\ +9 \\ \hline \end{array} \quad \begin{array}{r} 6 \\ +8 \\ \hline \end{array} \quad \begin{array}{r} 9 \\ +6 \\ \hline \end{array} \quad \begin{array}{r} 14 \\ -8 \\ \hline \end{array} \quad \begin{array}{r} 16 \\ -7 \\ \hline \end{array} \quad \begin{array}{r} 15 \\ -9 \\ \hline \end{array}$$

Lesson 9 Facts through 16

Add.

7 +8 15	6 +6	5 +9	8 +4	7 +7	6 +9
8 +8	9 +4	7 +6	8 +6	7 +9	8 +3
5 +6	8 +7	9 +7	8 +5	6 +8	9 +6

CHAPTER 9

Subtract.

14 − 7 7	15 − 6	16 − 7	13 − 8	14 − 5	12 − 9
12 − 6	14 − 6	11 − 9	16 − 8	11 − 3	15 − 9
15 − 7	13 − 9	16 − 9	14 − 9	15 − 8	14 − 8

Solve each problem.

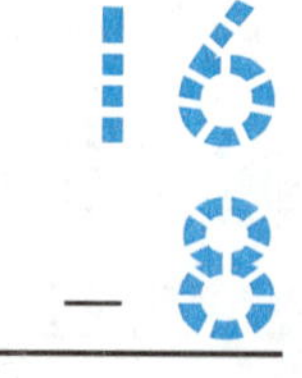

16	balloons in all
− 8	balloons broken
	balloons not broken

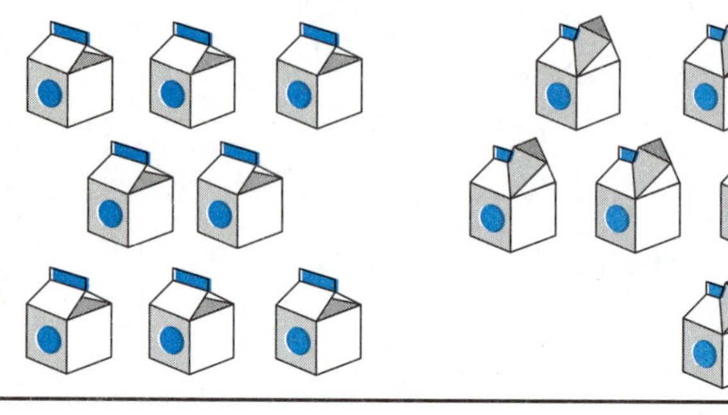

	cartons in all
−	cartons open
	cartons closed

	books in a box
+	books in a pile
	books in all

	glasses in all
−	glasses filled
	glasses empty

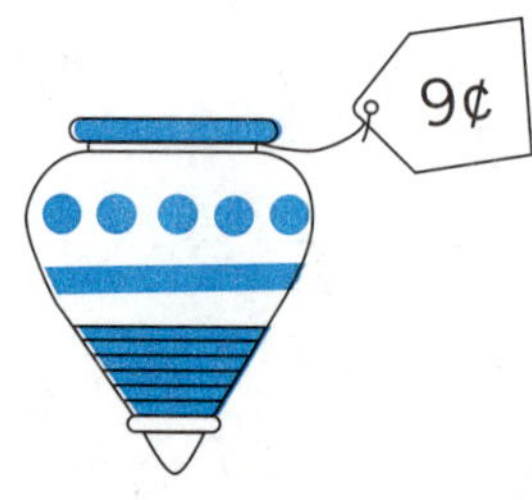

¢	for a yo-yo
+ ¢	for a top
¢	for both

Lesson 10 Facts through 18

Add or subtract.

$\begin{array}{r} 9 \\ +8 \\ \hline 17 \end{array}$ $\begin{array}{r} 8 \\ +9 \\ \hline \end{array}$ $\begin{array}{r} 17 \\ -8 \\ \hline 9 \end{array}$ $\begin{array}{r} 17 \\ -9 \\ \hline \end{array}$

$\begin{array}{r} 9 \\ +9 \\ \hline \end{array}$ $\begin{array}{r} 18 \\ -9 \\ \hline \end{array}$

$\begin{array}{r} 6 \\ +8 \\ \hline \end{array}$ $\begin{array}{r} 8 \\ +8 \\ \hline \end{array}$ $\begin{array}{r} 8 \\ +9 \\ \hline \end{array}$ $\begin{array}{r} 17 \\ -8 \\ \hline \end{array}$ $\begin{array}{r} 15 \\ -7 \\ \hline \end{array}$ $\begin{array}{r} 16 \\ -9 \\ \hline \end{array}$

$\begin{array}{r} 7 \\ +9 \\ \hline \end{array}$ $\begin{array}{r} 8 \\ +7 \\ \hline \end{array}$ $\begin{array}{r} 9 \\ +6 \\ \hline \end{array}$ $\begin{array}{r} 14 \\ -7 \\ \hline \end{array}$ $\begin{array}{r} 18 \\ -9 \\ \hline \end{array}$ $\begin{array}{r} 16 \\ -8 \\ \hline \end{array}$

$\begin{array}{r} 9 \\ +8 \\ \hline \end{array}$ $\begin{array}{r} 8 \\ +8 \\ \hline \end{array}$ $\begin{array}{r} 9 \\ +9 \\ \hline \end{array}$ $\begin{array}{r} 15 \\ -6 \\ \hline \end{array}$ $\begin{array}{r} 17 \\ -9 \\ \hline \end{array}$ $\begin{array}{r} 16 \\ -7 \\ \hline \end{array}$

CHAPTER 9

Solve each problem.

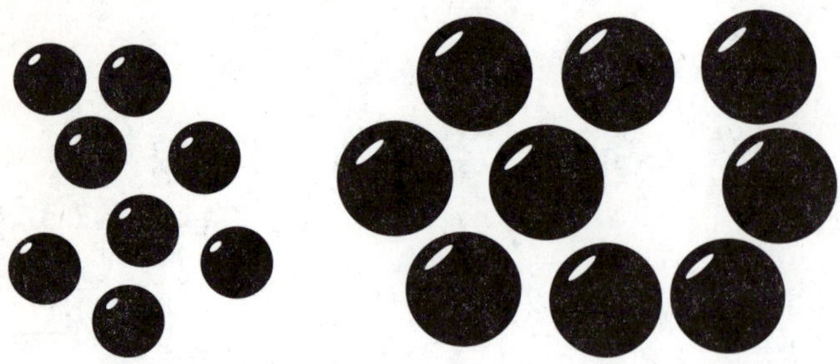

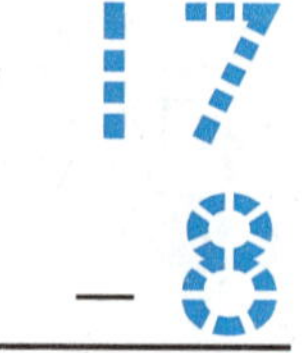

balls in all

small balls

large balls

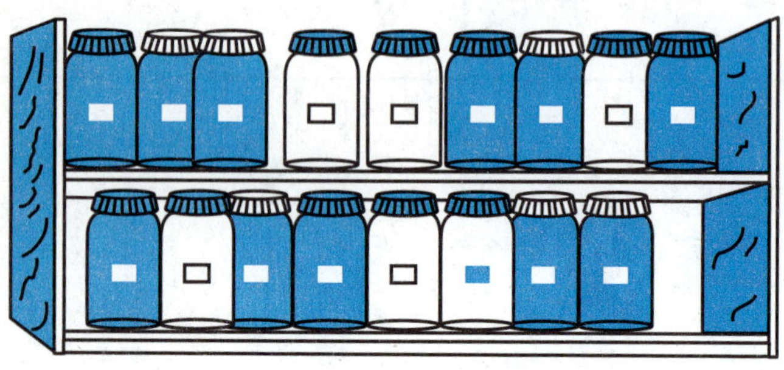

jars on top shelf

\+ ____ jars on bottom shelf

jars in all

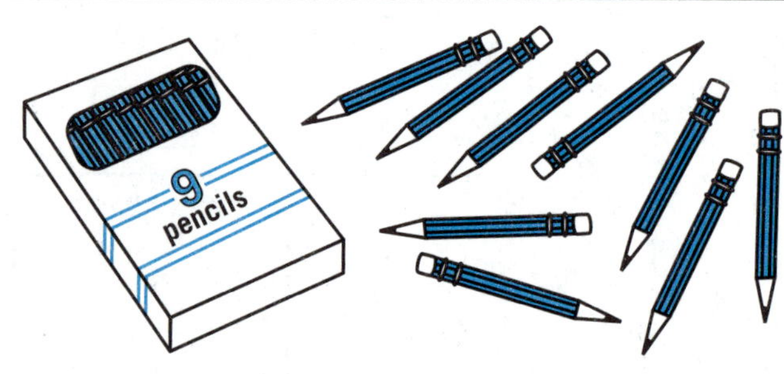

pencils in a box

\+ ____ more pencils

pencils in all

big stars

\+ ____ small stars

stars in all

bees in all

− ____ bees go away

bees stay

NAME ______________________

Lesson 11 Addition Facts through 18

Add.

5 +7 12	7 +5	2 +9	9 +2	5 +8	8 +5
6 +4	4 +6	9 +4	4 +9	6 +8	8 +6
7 +8	8 +7	4 +8	8 +4	9 +6	6 +9
7 +7	5 +6	6 +5	5 +9	9 + 5	8 + 8
5 +5	7 +6	6 +7	9 +8	8 +9	6 +6
7 +9	9 +7	4 +8	8 +3	7 +3	9 +9

Ring each fact for the number in the .

11 4 + 8 (9 + 2) (5 + 6) 6 + 6 (4 + 7) (8 + 3)	**12** 3 + 8 3 + 9 6 + 6 6 + 7 8 + 4 7 + 5 5 + 8 5 + 6
13 8 + 5 7 + 6 6 + 8 9 + 5 9 + 4 6 + 7 4 + 8	**14** 8 + 6 9 + 5 6 + 7 7 + 7 6 + 8 8 + 8
15 7 + 7 9 + 6 8 + 7 6 + 9 5 + 9 7 + 8	**16** 6 + 9 5 + 7 9 + 7 8 + 7 7 + 9 8 + 8
17 9 + 9 8 + 9 6 + 9 8 + 8 8 + 7 9 + 8	**18** 5 + 8 4 + 6 9 + 9 8 + 7 6 + 9

NAME ______________________

Lesson 12 Subtraction Facts through 18

Subtract.

15 − 7 = 8	15 − 8	13 − 9	13 − 4	12 − 8	12 − 4
14 − 6	14 − 8	11 − 7	11 − 4	10 − 7	10 − 3
15 − 6	15 − 9	14 − 9	14 − 5	17 − 8	17 − 9
16 − 8	16 − 7	16 − 9	13 − 5	13 − 8	14 − 7
12 − 6	11 − 3	11 − 8	13 − 7	13 − 6	10 − 5
10 − 8	10 − 2	12 − 5	12 − 7	18 − 9	15 − 7

CHAPTER 9

Lesson 12 Subtraction Facts through 18

How many subtraction facts do you know?

0	1	2	3	4	5	6	7	8	9
R	I	U	Y	E	K	O	V	W	N

Subtract. Then write the letter that goes with the answer.

$\begin{array}{r} 11 \\ -\ 8 \\ \hline 3 \end{array}$ $\begin{array}{r} 15 \\ -\ 9 \\ \hline \end{array}$ $\begin{array}{r} 10 \\ -\ 8 \\ \hline \end{array}$ $\begin{array}{r} 14 \\ -\ 9 \\ \hline \end{array}$ $\begin{array}{r} 16 \\ -\ 7 \\ \hline \end{array}$ $\begin{array}{r} 14 \\ -\ 8 \\ \hline \end{array}$ $\begin{array}{r} 15 \\ -\ 7 \\ \hline \end{array}$

$\begin{array}{r} 7 \\ -3 \\ \hline \end{array}$ $\begin{array}{r} 9 \\ -2 \\ \hline \end{array}$ $\begin{array}{r} 8 \\ -4 \\ \hline \end{array}$ $\begin{array}{r} 9 \\ -9 \\ \hline \end{array}$ $\begin{array}{r} 10 \\ -\ 7 \\ \hline \end{array}$ $\begin{array}{r} 11 \\ -\ 5 \\ \hline \end{array}$ $\begin{array}{r} 11 \\ -\ 2 \\ \hline \end{array}$ $\begin{array}{r} 13 \\ -\ 9 \\ \hline \end{array}$

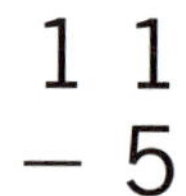

$\begin{array}{r} 12 \\ -\ 9 \\ \hline \end{array}$ $\begin{array}{r} 13 \\ -\ 7 \\ \hline \end{array}$ $\begin{array}{r} 11 \\ -\ 9 \\ \hline \end{array}$ $\begin{array}{r} 16 \\ -\ 8 \\ \hline \end{array}$ $\begin{array}{r} 10 \\ -\ 9 \\ \hline \end{array}$ $\begin{array}{r} 18 \\ -\ 9 \\ \hline \end{array}$

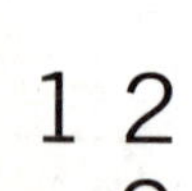

NAME ______________________

CHAPTER 9 PRACTICE TEST

Addition and Subtraction (facts through 18)

Add.

9	5	8	7	4	5
+8	+7	+6	+5	+9	+6

7	8	5	9	7	6
+6	+8	+9	+9	+8	+6

4	9	7	8	9	7
+7	+2	+7	+4	+6	+9

Subtract.

13	14	15	11	12	16
−6	−7	−6	−3	−8	−9

13	12	11	17	18	15
−8	−9	−5	−8	−9	−7

14	11	12	14	11	17
−8	−7	−6	−5	−9	−9

CHAPTER 9

Problem Solving

Solve each problem.

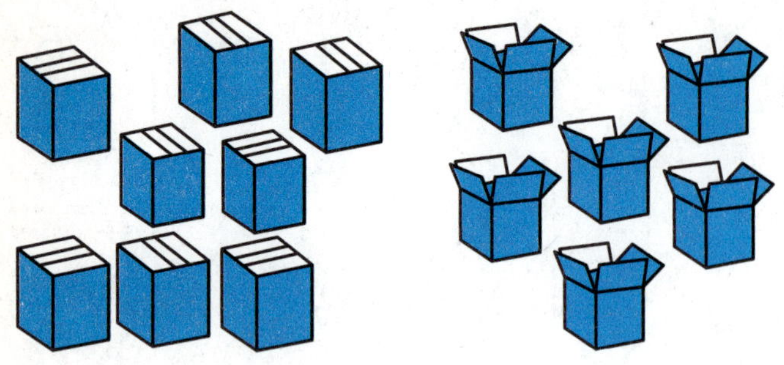

closed boxes

\+ ____ open boxes

boxes in all

candles in all

− ____ blue candles

white candles

jars in a box

\+ ____ other jars

jars in all

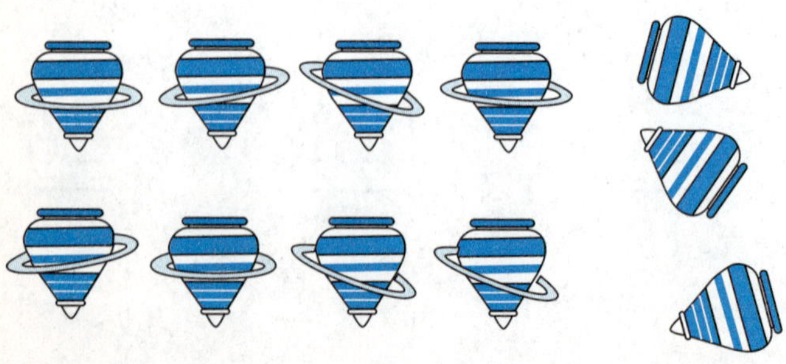

tops in all

− ____ tops not spinning

tops spinning

NAME ______________________

MID-TEST Chapters 1–4

Ring the numeral.

0 1 2 3	4 5 6 7	7 8 9 10
2¢ 5¢ 6¢	1¢ 5¢ 10¢	4¢ 8¢ 9¢

Ring the  in the place named.

fourth

third

   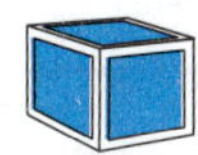 

Add.

5	3	4	3	2	3
+5	+6	+3	+1	+6	+3

Subtract.

9	7	10	9	10	5
−7	−5	−4	−9	−9	−2

MID-TEST CH. 1-4

GO

Add or subtract. STOP! Watch the + and − signs!

$\begin{array}{r} 4 \\ +6 \\ \hline \end{array}$ $\begin{array}{r} 6 \\ -1 \\ \hline \end{array}$ $\begin{array}{r} 4 \\ +4 \\ \hline \end{array}$ $\begin{array}{r} 10 \\ -5 \\ \hline \end{array}$ $\begin{array}{r} 1 \\ +8 \\ \hline \end{array}$ $\begin{array}{r} 8 \\ -7 \\ \hline \end{array}$

Solve each problem.

6¢

3¢

1¢

I buy  ¢

and TICKET + ¢

I spent ¢

I have 9¢

I buy  − ¢

I have left ¢

Matt had 7  .

He gave 2 away.

How many does he have now?

There are 6 blue .

There are 3 white .

How many are there in all?

STOP

FINAL TEST Chapters 1–9

Tell how many cents.

____ ¢

____ ¢

____ ¢

Name the numbers in order.

56, 57, 58, ____, ____, ____, ____

Write the time for each clock.

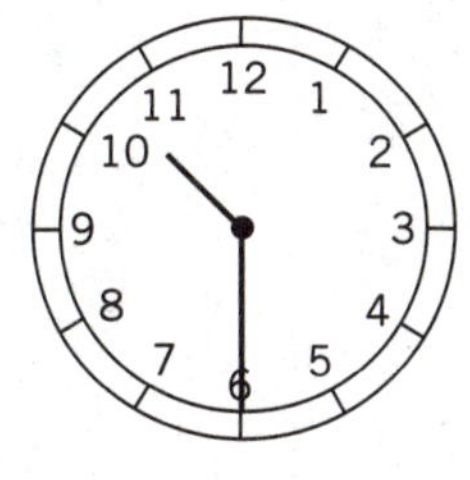

____:____

____:____

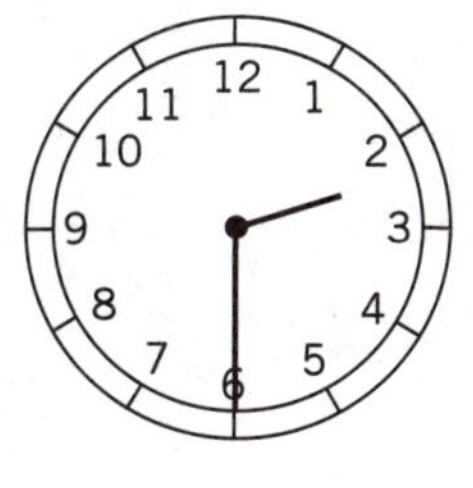

____:____

Automobiles

How many  ? 

How many  ? 

How many  ? 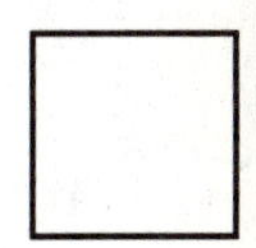

FINAL TEST CH. 1–9

How long is each object?

Use a centimeter ruler.

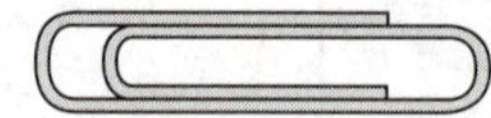

_____ centimeters

_____ centimeters

Use an inch ruler.

_____ inches

_____ inches

Add.

3 + 1	6 + 4	4 + 5	0 + 7	4 + 4
20 + 14	71 + 3	21 + 25	42 + 15	12 + 67

Subtract.

5 − 3	7 − 1	10 − 4	9 − 7	8 − 5
68 − 48	72 − 50	39 − 4	84 − 41	57 − 52

GO

Add or subtract. STOP! Watch the + and − signs.

$\begin{array}{r} 10 \\ -\ 2 \\ \hline \end{array}$ $\begin{array}{r} 5 \\ +5 \\ \hline \end{array}$ $\begin{array}{r} 7 \\ +\ 2 \\ \hline \end{array}$ $\begin{array}{r} 7 \\ -3 \\ \hline \end{array}$ $\begin{array}{r} 10 \\ -7 \\ \hline \end{array}$

$\begin{array}{r} 51 \\ +27 \\ \hline \end{array}$ $\begin{array}{r} 69 \\ -10 \\ \hline \end{array}$ $\begin{array}{r} 43 \\ +\ 2 \\ \hline \end{array}$ $\begin{array}{r} 87 \\ -17 \\ \hline \end{array}$ $\begin{array}{r} 76 \\ -62 \\ \hline \end{array}$

Add.

$\begin{array}{r} 3 \\ +8 \\ \hline \end{array}$ $\begin{array}{r} 7 \\ +7 \\ \hline \end{array}$ $\begin{array}{r} 8 \\ +\ 4 \\ \hline \end{array}$ $\begin{array}{r} 9 \\ +9 \\ \hline \end{array}$ $\begin{array}{r} 8 \\ +8 \\ \hline \end{array}$

$\begin{array}{r} 8 \\ +7 \\ \hline \end{array}$ $\begin{array}{r} 4 \\ +9 \\ \hline \end{array}$ $\begin{array}{r} 7 \\ +\ 6 \\ \hline \end{array}$ $\begin{array}{r} 6 \\ +8 \\ \hline \end{array}$ $\begin{array}{r} 8 \\ +9 \\ \hline \end{array}$

Subtract.

$\begin{array}{r} 12 \\ -\ 6 \\ \hline \end{array}$ $\begin{array}{r} 14 \\ -\ 7 \\ \hline \end{array}$ $\begin{array}{r} 11 \\ -\ 8 \\ \hline \end{array}$ $\begin{array}{r} 15 \\ -\ 9 \\ \hline \end{array}$ $\begin{array}{r} 17 \\ -\ 8 \\ \hline \end{array}$

$\begin{array}{r} 13 \\ -\ 9 \\ \hline \end{array}$ $\begin{array}{r} 15 \\ -\ 8 \\ \hline \end{array}$ $\begin{array}{r} 16 \\ -\ 7 \\ \hline \end{array}$ $\begin{array}{r} 14 \\ -\ 6 \\ \hline \end{array}$ $\begin{array}{r} 18 \\ -\ 9 \\ \hline \end{array}$

GO

Ring the shape.

square  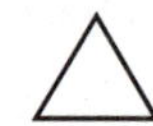 

cylinder 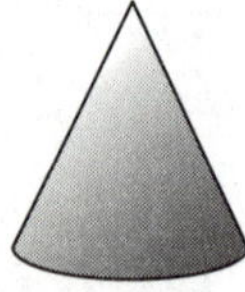  

Solve each problem.

Sara has 2  .

Ryan has 6 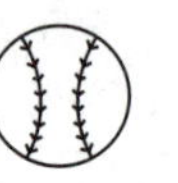 .

+

How many do they have in all?

You had 38¢. 

You spent 23¢. 

How much do you have left? 

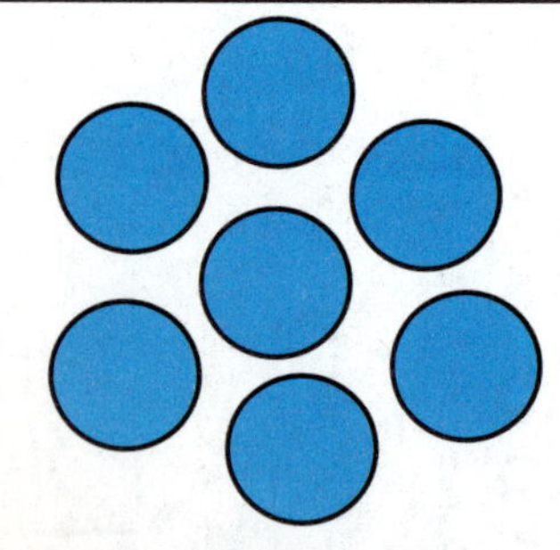 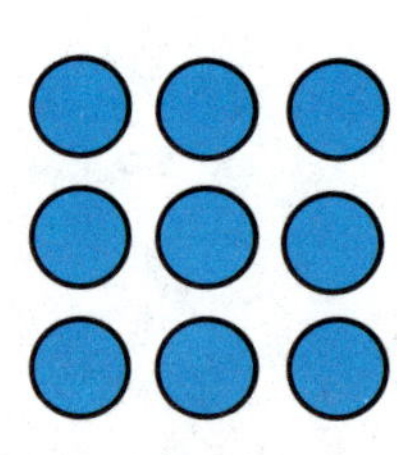

large balls

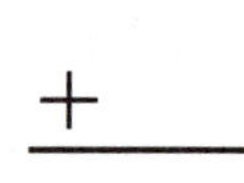 small balls

balls in all

cars in all

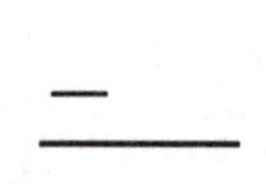 black cars

white cars

CHAPTER 1 CUMULATIVE REVIEW

Work each problem.
Find the correct answer.
Mark the space for the answer.

Part 1 Concepts

Tell how many.

1.

A 5
B 6
C 7
D 8

2.

F 6
G 7
H 8
J 9

3.

A 3
B 4
C 5
D 6

Part 2 Computation

What numeral goes in the ☐ to finish the pattern?

4. 1 2 ☐ 4 5

F 6 H 3
G 7 J 8

5. 4 5 6 ☐ 8

A 7 C 9
B 8 D 10

Part 3 Applications

6. Name the place of the ringed box.

F first
G second
H third
J fourth

CUMULATIVE REVIEW

ANSWER ROW
1 Ⓐ Ⓑ Ⓒ Ⓓ 3 Ⓐ Ⓑ Ⓒ Ⓓ 5 Ⓐ Ⓑ Ⓒ Ⓓ
2 Ⓕ Ⓖ Ⓗ Ⓙ 4 Ⓕ Ⓖ Ⓗ Ⓙ 6 Ⓕ Ⓖ Ⓗ Ⓙ

NAME ______________________

CHAPTER 2 CUMULATIVE REVIEW

Work each problem.
Find the correct answer.
Mark the space for the answer.

Part 1 Concepts

1. Tell how many.

A 5
B 6
C 7
D 8

□ □ □ □ □ □ □ □

2. Finish the pattern.
What number goes in the □?

2 3 □ 5 6

F 4 H 6
G 5 J 7

Part 2 Computation

3. Add.

$$\begin{array}{r} 3 \\ +1 \\ \hline \end{array}$$

A 1
B 2
C 3
D 4

4. Subtract.

$$\begin{array}{r} 4 \\ -3 \\ \hline \end{array}$$

F 1
G 2
H 3
J 4

Part 3 Applications

5. There are 5 .

2 fly away.

How many are left?

A 2 C 6
B 3 D 7

6. Bill has 2 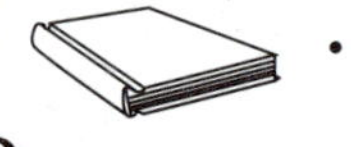.

Ann has 2 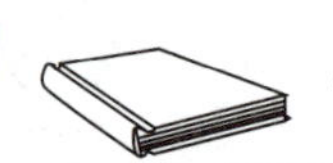.

How many in all?

F 4 H 3
G 5 J 2

ANSWER ROW
1 Ⓐ Ⓑ Ⓒ Ⓓ 3 Ⓐ Ⓑ Ⓒ Ⓓ 5 Ⓐ Ⓑ Ⓒ Ⓓ
2 Ⓕ Ⓖ Ⓗ Ⓙ 4 Ⓕ Ⓖ Ⓗ Ⓙ 6 Ⓕ Ⓖ Ⓗ Ⓙ

CHAPTER 3 CUMULATIVE REVIEW

Work each problem.
Find the correct answer.
Mark the space for the answer.

Part 1 Concepts

1. Tell how many.

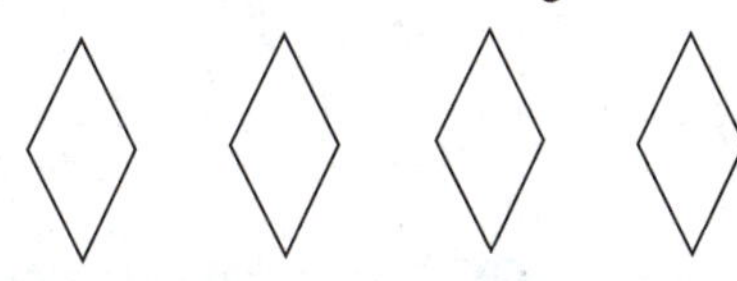

A 3 C 5
B 4 D 6

2. Name the place of the ringed ball.

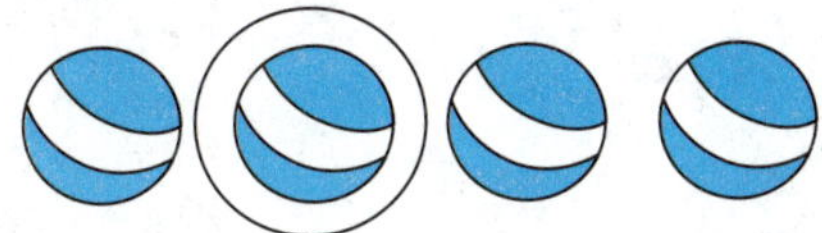

F first
G second
H third
J fourth

Part 2 Computation

3. Add.

$$\begin{array}{r} 4 \\ +3 \\ \hline \end{array}$$

A 6 C 8
B 7 D 9

4. Subtract.

$$\begin{array}{r} 8 \\ -6 \\ \hline \end{array}$$

F 1 H 3
G 2 J 4

Part 3 Applications

5. There are 4 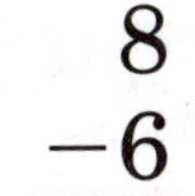.

There are 3 .

How many in all?

A 4
B 5
C 6
D 7

6. There are 3 .

5 more come.

How many in all?

F 6
G 7
H 2
J 8

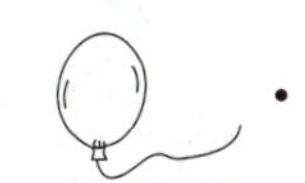

CUMULATIVE REVIEW

ANSWER ROW

1 Ⓐ Ⓑ Ⓒ Ⓓ 3 Ⓐ Ⓑ Ⓒ Ⓓ 5 Ⓐ Ⓑ Ⓒ Ⓓ
2 Ⓕ Ⓖ Ⓗ Ⓙ 4 Ⓕ Ⓖ Ⓗ Ⓙ 6 Ⓕ Ⓖ Ⓗ Ⓙ

NAME ______________________

CHAPTER 4 CUMULATIVE REVIEW

Work each problem.
Find the correct answer.
Mark the space for the answer.

Part 1 Concepts

1. Tell how many.

A four
B five
C six
D seven

2. Count forward. What numeral goes in the ☐?

1 2 3 ☐ 5 6

F 3 H 5
G 4 J 6

Part 2 Computation

3. Add.

$$\begin{array}{r} 6 \\ +4 \\ \hline \end{array}$$

A 7
B 8
C 9
D 10

4. Subtract.

$$\begin{array}{r} 9 \\ -2 \\ \hline \end{array}$$

F 5 H 7
G 6 J 8

Part 3 Applications

5. There were 9  .

6 were eaten.

How many were not eaten?

A 2
B 3
C 4
D 5

6. I have 5  .

I find 4 more.

Now how many in all?

F 7 H 9
G 8 J 10

ANSWER ROW

1 Ⓐ Ⓑ Ⓒ Ⓓ 3 Ⓐ Ⓑ Ⓒ Ⓓ 5 Ⓐ Ⓑ Ⓒ Ⓓ
2 Ⓕ Ⓖ Ⓗ Ⓙ 4 Ⓕ Ⓖ Ⓗ Ⓙ 6 Ⓕ Ⓖ Ⓗ Ⓙ

NAME ______________________

CHAPTER 5 CUMULATIVE REVIEW

Work each problem.
Find the correct answer.
Mark the space for the answer.

Part 1 Concepts

Complete.

1. 1 ten 6 ones = ______

A 16
B 15
C 51
D 61

2. 4 tens 8 ones = ______

F 47
G 48
H 78
J 87

Part 2 Computation

3. Add.

$$\begin{array}{r} 2 \\ +8 \\ \hline \end{array}$$

A 7
B 8
C 9
D 10

4. Subtract.

$$\begin{array}{r} 9 \\ -6 \\ \hline \end{array}$$

F 2
G 3
H 4
J 5

Part 3 Applications

5. There were 6 .

4 swam away.

How many were left?

A 2
B 3
C 4
D 5

6. There are 8 .

Nick finds 2 more.

Now how many in all?

F 7
G 8
H 9
J 10

CUMULATIVE REVIEW

ANSWER ROW
1 Ⓐ Ⓑ Ⓒ Ⓓ
2 Ⓕ Ⓖ Ⓗ Ⓙ
3 Ⓐ Ⓑ Ⓒ Ⓓ
4 Ⓕ Ⓖ Ⓗ Ⓙ
5 Ⓐ Ⓑ Ⓒ Ⓓ
6 Ⓕ Ⓖ Ⓗ Ⓙ

NAME ______________________

CHAPTER 6 CUMULATIVE REVIEW

Work each problem.
Find the correct answer.
Mark the space for the answer.

Part 1 Concepts

1. Complete.

5 tens 2 ones = ______

A 25 C 52
B 26 D 62

2. Count by 2s. What number goes in the ___?

6, 8, __, 12, 14

F 12 H 4
G 10 J 16

Part 2 Computation

3. Add.

$$\begin{array}{r} 7 \\ +2 \\ \hline \end{array}$$

A 7
B 8
C 9
D 10

4. Subtract.

$$\begin{array}{r} 10 \\ -3 \\ \hline \end{array}$$

F 5
G 6
H 7
J 8

Part 3 Applications

5. What time is on the clock?

A 4:00
B 6:00
C 4:30
D 5:30

6. How long is the pencil?

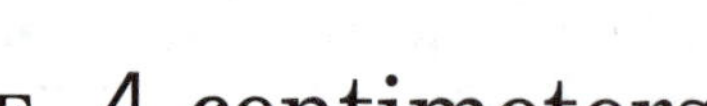

F 4 centimeters
G 5 centimeters
H 6 centimeters
J 7 centimeters

ANSWER ROW
1 Ⓐ Ⓑ Ⓒ Ⓓ 3 Ⓐ Ⓑ Ⓒ Ⓓ 5 Ⓐ Ⓑ Ⓒ Ⓓ
2 Ⓕ Ⓖ Ⓗ Ⓙ 4 Ⓕ Ⓖ Ⓗ Ⓙ 6 Ⓕ Ⓖ Ⓗ Ⓙ

NAME ______________________

CHAPTER 7 CUMULATIVE REVIEW

Work each problem.
Find the correct answer.
Mark the space for the answer.

Part 1 Concepts

1. Which number is the greatest?

34   40   8   22

A 34   C 8
B 40   D 22

2. Count by 5s. What number goes in the ___?

5, 10, 15, __, 25

F 15
G 20
H 25
J 30

Part 2 Computation

3. Add.

$$\begin{array}{r} 6 \\ +2 \\ \hline \end{array}$$

A 6
B 7
C 8
D 9

4. Subtract.

$$\begin{array}{r} 9 \\ -4 \\ \hline \end{array}$$

F 4
G 5
H 6
J 7

Part 3 Applications

5. What time is on the clock?

A two-thirty
B three-thirty
C four-thirty
D five-thirty

6. Which shape is shown?

F circle
G rectangle
H square
J triangle

ANSWER ROW

1 Ⓐ Ⓑ Ⓒ Ⓓ   3 Ⓐ Ⓑ Ⓒ Ⓓ   5 Ⓐ Ⓑ Ⓒ Ⓓ
2 Ⓕ Ⓖ Ⓗ Ⓙ   4 Ⓕ Ⓖ Ⓗ Ⓙ   6 Ⓕ Ⓖ Ⓗ Ⓙ

NAME ______________________

Work each problem.
Find the correct answer.
Mark the space for the answer.

Part 1 Concepts

1. Tell how many.

A 15
B 16
C 17
D 18

2. Count by 10s. What number goes in the ___?

20, 30, 40, __, 60

F 30
G 40
H 50
J 60

Part 2 Computation

3. Add.

$$\begin{array}{r} 40 \\ +32 \\ \hline \end{array}$$

A 62
B 72
C 82
D 92

4. Subtract.

$$\begin{array}{r} 39 \\ -24 \\ \hline \end{array}$$

F 5
G 15
H 25
J 35

Part 3 Applications

5. Which shape is shown?

A cone
B cube
C cylinder
D sphere

6. How long is the crayon?

CRAYON

F 4 inches
G 5 inches
H 6 inches
J 7 inches

STOP

ANSWER ROW
1 Ⓐ Ⓑ Ⓒ Ⓓ 3 Ⓐ Ⓑ Ⓒ Ⓓ 5 Ⓐ Ⓑ Ⓒ Ⓓ
2 Ⓕ Ⓖ Ⓗ Ⓙ 4 Ⓕ Ⓖ Ⓗ Ⓙ 6 Ⓕ Ⓖ Ⓗ Ⓙ

NAME ______________________

Work each problem.
Find the correct answer.
Mark the space for the answer.

Part 1 Concepts

1. Tell how many.

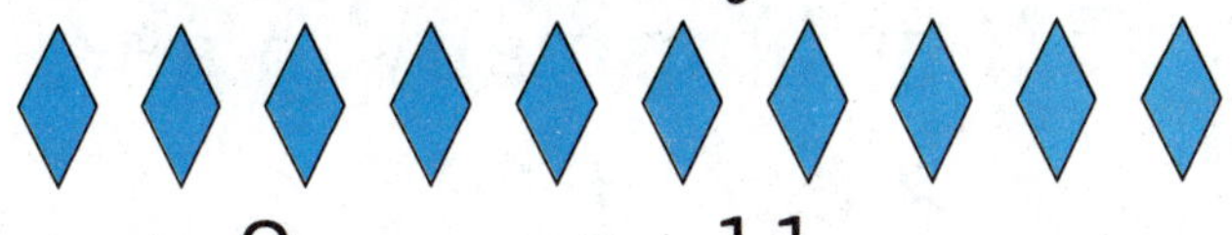

A 9 C 11
B 10 D 12

2. Which number is the greatest?

5 2 12 10

F 5
G 2
H 12
J 10

3. Count by 2s. What number goes in the ___?

24, 26, __, 30, 32

A 27
B 28
C 29
D 30

Part 2 Computation

4. Add.

$$\begin{array}{r} 7 \\ +6 \\ \hline \end{array}$$

F 12
G 13
H 14
J 15

5. Subtract.

$$\begin{array}{r} 15 \\ -8 \\ \hline \end{array}$$

A 5
B 6
C 7
D 8

6. Add.

$$\begin{array}{r} 9 \\ +6 \\ \hline \end{array}$$

F 12
G 13
H 14
J 15

GO

ANSWER ROW

1 Ⓐ Ⓑ Ⓒ Ⓓ 3 Ⓐ Ⓑ Ⓒ Ⓓ 5 Ⓐ Ⓑ Ⓒ Ⓓ
2 Ⓕ Ⓖ Ⓗ Ⓙ 4 Ⓕ Ⓖ Ⓗ Ⓙ 6 Ⓕ Ⓖ Ⓗ Ⓙ

7. Subtract.

$$\begin{array}{r} 12 \\ -3 \\ \hline \end{array}$$

A 7
B 8
C 9
D 10

8. Add.

$$\begin{array}{r} 54 \\ +12 \\ \hline \end{array}$$

F 66
G 56
H 46
J 36

9. Subtract.

$$\begin{array}{r} 49 \\ -13 \\ \hline \end{array}$$

A 16
B 26
C 36
D 46

Part 3 Applications

10. There are 6  .

I find 8 more.

Now how many in all?

F 7 H 16
G 14 J 18

11. Which shape is shown?

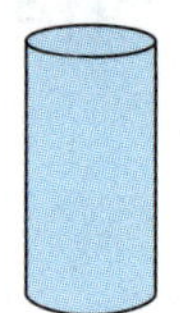

A cone
B cube
C cylinder
D rectangular solid

12. How long is the ribbon?

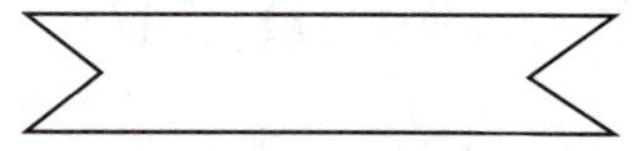

F 2 centimeters
G 3 centimeters
H 4 centimeters
J 5 centimeters

STOP

ANSWER ROW
7 Ⓐ Ⓑ Ⓒ Ⓓ 9 Ⓐ Ⓑ Ⓒ Ⓓ 11 Ⓐ Ⓑ Ⓒ Ⓓ
8 Ⓕ Ⓖ Ⓗ Ⓙ 10 Ⓕ Ⓖ Ⓗ Ⓙ 12 Ⓕ Ⓖ Ⓗ Ⓙ